Caribbean Recipes for Schools

Alison J. Rigby, BSc.

MACMILLAN
CARIBBEAN

Copyright © Alison J. Rigby 1987

All rights reserved. No reproduction, copy or transmission
of this publication may be made without written permission.
No paragraph of this publication may be reproduced, copied
or transmitted save with written permission or in accordance
with the provisions of the Copyright Act 1956 (as amended).
Any person who does any unauthorised act in relation to
this publication may be liable to criminal prosecution and
civil claims for damages.

First published 1987

Published by *Macmillan Publishers Ltd*
London and Basingstoke
Associated companies and representatives in Accra.
Auckland, Delhi, Dublin, Gaborone, Hamburg, Harare,
Hong Kong, Kuala Lumpur, Lagos, Manzini, Melbourne,
Mexico City, Nairobi, New York, Singapore, Tokyo.

ISBN 0-333-44682-8

Printed in Hong Kong

British Library Cataloguing in Publication Data
Rigby, Alison
  Caribbean recipes for schools.
  1. Cookery, Caribbean
  I. Title
  641.59'1821   TX716.A1
ISBN 0-333-44682-8

CONTENTS

Acknowledgements	iv
Introduction	1
Conversion tables	3
The six food groups	5
Hygiene in the kitchen	7
Safety in the kitchen	9
Using recipes	10
Weighing and measuring	11
Food group 1: Staples	13
Food group 2: Legumes	23
Food group 3: Dark green leafy/yellow vegetables	34
Food group 4: Food from animals	44
Food group 5: Fruits	68
Food group 6: Fats and oils	79
Cookery terms	90
Methods	92
Meal planning	94
Menu ideas	96
Recipe index	98

ACKNOWLEDGEMENTS

The author would like to thank the Ministry of Education, Nassau, Bahamas and the Caribbean Food and Nutrition Institute as sources of information.

INTRODUCTION

This book has been compiled for use in conjunction with the other books in the *Macmillan Caribbean Home Economics (1−4)* series. It is intended for use in schools by both teacher and students (aged 11 to 16) who are following a Food and Nutrition programme as part of their Home Economics course.

There is a strong emphasis on healthy eating, based on the nutritional principles of planning meals around the six food groups. The recipes which comprise the main part of the book make it easy to learn the foods within each group and to plan nutritious meals. These sound nutritional guidelines can be followed up in more depth work in nutrition, from books 1, 2 and 3 of *Caribbean Home Economics*.

As well as being nutritionally balanced, healthy meals should also have an acceptable taste and make economical use of ingredients. Other countries have a large influence on the Caribbean's eating habits, which is reflected in the wide range of packet convenience foods available. However, as far as possible the rich supply of nutritious fresh fish and Caribbean fruit and vegetables have been used. New and imaginative recipe ideas have been included in addition to some traditional dishes which make use of many practical skills.

For examination purposes (including the G.C.E. 'O' level practical examination) this book includes a wide range of methods of cookery and cookery terms. Throughout the book there is work on sauces, soups and salad dressings; preservation; yeast mixtures; cake-making; pastry-making; batters; meat and fish cookery; use of gelatine and vegetable preparation. Special equipment and correct weighing and measuring with recipes written in three measurements (ounces, grammes and cup sizes) have been carefully tested for use throughout the Caribbean.

Use the recipes in connection with each other: for example, plan work on the uses of eggs in cookery, sauce-making or pastry-making. Vary the recipes according to individual taste: for example, decrease the amount of sugar used; try wholewheat instead of all-purpose flour and add locally-grown food or a particular flavouring or spice. Many of the islands in the Caribbean have different names for their own produce which can be substituted in the recipe. An awareness of some of the problems associated with working in schools in the Caribbean makes this a valuable guide.

Above all Home Economics should be fun and this little book will help everyone to enjoy the food preparation section.

CONVERSION TABLES

These tables are used in the recipes as equivalent weights and measures.

Weight

Imperial ounces	Metric grammes	Imperial ounces	Metric grammes
½ oz	15 g	8 oz (½ lb)	225 g
1 oz	25 g	9 oz	250 g (¼ kg)
1½ oz	40 g	10 oz	275 g
2 oz	50 g	11 oz	300 g
3 oz	75 g	12 oz (¾ lb)	350 g
4 oz (¼ lb)	100 g	13 oz	375 g
5 oz	125 g	14 oz	400 g
6 oz	150 g	15 oz	425 g
7 oz	175 g	16 oz (1 lb)	450 g

Volume

Imperial fluid ounces	Metric millilitres
1 fl oz	25 ml
4 fl oz (¼ pt)	125 ml
8 fl oz (½ pt)	250 ml
12 fl oz (¾ pt)	350 ml
16 fl oz (1 pt)	500 ml (½ litre)
32 fl oz (2 pt)	1000 ml (1 litre)

* All the measurements in this book refer to an American ½ pint measure which holds 8 fluid ounces. A British ½ pint measure holds 10 fluid ounces.

Spoon sizes
1 teaspoon = 5 ml   1 tablespoon = 15 ml

Oven temperatures

	Fahrenheit °F	Celsius °C	Gas Mark Regulo
Very slow	225	110	$\frac{1}{4}$
	250	130	$\frac{1}{2}$
Slow	275	140	1
	300	150	2
	325	170	3
Warm	350	180	4
Moderate	375	190	5
Fairly hot	400	200	6
	425	220	7
Hot	450	230	8
Very hot	475	240	9

THE SIX FOOD GROUPS

Try to choose at least one food from as many groups as possible at each meal.

1 Staples

(a) Cereals,
for example: Wheat – breakfast cereal, bread, pasta (macaroni and spaghetti).
Maize – cornbread, cornmeal, cornstarch, grits and fresh or canned sweetcorn.
Oats – breakfast cereal (porridge) and oatmeal.
Rice – rice pudding and packet rice for meals.
(b) Starchy fruit and vegetables,
for example: breadfruit, plantain (green banana), cassava, sweet potato, Irish potato, yam, tannia, dasheen, eddoe and coco.

2 Legumes

for example: pigeon (gungo) peas, black-eye peas, split peas, chick peas, soya beans, haricot beans, broad beans, kidney beans, lentils and peanuts (ground nuts).

3 Dark green leafy/yellow vegetables

for example: callaloo, spinach, pak choi, dasheen leaves, broccoli, christophene, pumpkin, calabaza, squash, egg-plant, carrot, tomato and okra.

4 Food from animals

for example: meat, poultry (chicken and turkey), fish, eggs, milk, cheese and yogurt.

5 Fruits

for example: guava, papaya (pawpaw), pineapple, mango, soursop, cherry, sugar apple, orange, grapefruit, lime, sapodilla and banana.

6 Fats and oils

for example: margarine, butter, shortening, cooking oil, salt pork, suet, coconut, avocado pear and salad dressing.

Healthy eating guidelines

Eating sensibly keeps us fit and well. Food is necessary for the growth and repair of our bodies and to provide the energy for keeping the body warm and for all activities.

Combine foods from the six food groups for good nutrition. If you include foods from each of these groups every day, you will be making interesting balanced meals.

A balanced diet or meal is one which provides the body with all the essential nutrients. Nutrients are the chemical substances found in food which enable the body to grow and function properly. These are **protein, carbohydrate, fat, vitamins** and **minerals**. All the necessary nutrients will be provided by a diet which includes foods from all six food groups. Additionally **fibre** and **water**, although not called nutrients, are both needed for good health.

All six food groups are required to work together. Eating foods from one group alone cannot supply all the needs of your body. Some of the recipes in the book include foods from more than one of the six food groups, for example *Macaroni Cheese* is found in the food from the animal group and contains a staple food. 'Family pot' type meals, such as chowders, soups and casseroles will also contain a variety of foods.

Individuals who cannot eat meat can still maintain a healthy diet. In the case of a strict vegetarian, an extra serving from the legume group should be included. Everyone can increase the amount of fibre in his or her diet with fresh fruit and vegetables and use alternative protein sources, such as cereals, peas and beans.

Sugar does not appear in any of the food groups, as it provides us with energy and nothing else. Foods with a high sugar content should be eaten in moderate amounts and not as a substitute for sensible eating. It is also preferable to limit the amount of salt used in cooking and, if necessary, add more to taste when the food is served.

HYGIENE IN THE KITCHEN

Hygiene is the practice of cleanliness, for the maintenance of good health.

Kitchen hygiene ensures that food does not become infected with harmful bacteria which can be transmitted by flies, dust, humans, pets (dogs and cats) and vermin (rats, mice and roaches). Ensure that you follow these rules when preparing food in the kitchen:

A. Personal hygiene

1. Wash your hands before cooking and always after going to the toilet.
2. Wear a clean apron.
3. Keep your hair tied out of the way of food.
4. Avoid wearing nail polish and jewellery, especially rings.
5. Never lick your fingers or utensils and then put them into food.
6. Wash your hands after using a handkerchief when touching food.
7. Do not handle food if you have a bad cold or sickness.
8. Cover all cuts with a clean bandage.

B. Kitchen cleanliness

1. Keep all utensils, equipment and work surfaces clean.
2. Wash up in water with detergent as you work and avoid chipped utensils.
3. Keep all cloths used in the kitchen clean.
4. Wrap all scraps of food in paper and place them in a covered bin.
5. Empty garbage bins frequently and wash them out with disinfectant.
6. Sweep up any spilt crumbs and mop floors with water and disinfectant.

C. Food storage

KEEP FOOD COOL, CLEAN AND COVERED.

Most fresh (perishable) foods should be stored in the refrigerator, in a clean container or on a plate and covered with foil wrap or a plastic bag. If cooked food must be stored, cool it rapidly first. When reheating food, heat it thoroughly and do not reheat more than once.

SAFETY IN THE KITCHEN

Safety is avoiding risk or danger, so that accidents do not happen.

The major kitchen accident groups include: burns and scalds, cuts and scrapes, falls and poisoning. In order to ensure that the kitchen is as safe a place as possible careful rules should be followed:

1. Keep cooking pot handles turned inwards on top of the stove.
2. Use an oven cloth or gloves when using the oven and when lifting hot dishes.
3. Avoid touching electrical switches and plugs with wet hands.
4. Always stand guard when frying foods and never leave a deep fat pan unattended.
5. Keep kitchen cloths, towels and curtains away from a working stove.
6. Wipe up any spills immediately and do not leave chairs, toys and rugs where someone might fall over them.
7. Keep sharp equipment, plastic bags, household cleaners, matches and medicines well out of reach of young children.
8. Keep a First Aid kit at hand which includes: scissors, sterile bandages and dressings, antiseptic cream or lotion, cotton wool and tweezers.

USING RECIPES

A recipe is the complete description of how to make any of the dishes used, for example in this book. Each recipe includes four main points:
1. Ingredients – including both local and packaged foods.
2. Amount of ingredients – measured in ounces, grammes and cup sizes.
3. Method – instructions on using the ingredients.
4. Baking time and temperature – for gas and electric cookers.

The fuels used for cooking vary throughout the Caribbean and the recipes can also be adapted for use with oil or solid fuel cookers, fireplace (clay or brick) or charcoal fires and iron pots. Pre-heat the oven 10 to 15 minutes before it is required and move the shelves while still cold into the correct position. When using an electric cooker check that the main switch is on and turn on the oven dial, which usually has a light to show that it is on. For a gas cooker, turn on the gas and light using a match (or self-ignition) and adjust the temperature with the 'regulo' switch.

In this book the top of the cooker is referred to as the stove (hob), inside as the oven (baker); the grill or broiler is usually situated at eye level or below the hob. Usually the top part of the oven is hotter than the lower shelves. The temperature of the ovens may vary and should be tested with a thermometer if they are 'hot' (cooking too quickly) or 'slow' (taking a long time to cook).

The recipes can be reproduced in a short amount of time and will serve approximately four people. When using the equipment called for in the recipes, try to use the correct utensils wherever possible to make your cooking quicker, easier and safer. If modern equipment is not at hand substitute what is available, for example use a sharp knife instead of a vegetable peeler. Watch out for the following abbreviations used in the recipes:

tsp = teaspoon   oz = ounce   pt = pint
tblsp = tablespoon   lb = pound   C = cup

WEIGHING AND MEASURING

For good results when following the recipes in this book the ingredients should be carefully weighed or measured.

Solid food, e.g. flour and sugar, can be weighed using scales, or measured with spoons or cups. The types of scales which may be available in the kitchen are: the traditional balance scale, which has weights at one end balanced with the ingredients at the other or other scales with the figures written horizontally or on a dial face. When using this type of scale always check that the pointer is at O when the scale pan is empty. It is usually easier to look at a packet or tub of fat for the weight and cut pieces off until you have the right amount. Net weight refers to the weight of the ingredients alone without the container.

A set of measuring spoons is a good idea. Generally 'spoonful' sizes in the book refer to a rounded spoonful. A level teaspoon or tablespoon means that the ingredients are levelled off across the rim of the spoon using your finger or a knife.

Liquids, e.g. milk and fruit juice, are also measured with spoons or cups. Remember that a standard American cup measurement, which holds $\frac{1}{2}$ pint or 8 fluid ounces, is used in all the measurements. Use the following table to assist you when measuring foods in cup sizes. As you will see the capacity will vary according to the type of food being measured.

Kitchen calculations

Volume		Imperial	Metric
1 cup liquid	=	$\frac{1}{2}$ pt	250 ml
1 cup sugar	=	8 oz	225 g
1 cup butter or margarine	=	8 oz	225 g
1 cup rice – dried	=	8 oz	225 g
rice – cooked	=	4 oz	100 g
1 cup beans – dried	=	8 oz	225 g
beans – cooked	=	4 oz	100 g
1 cup cornmeal/cornstarch	=	5 oz	125 g

Volume		Imperial	Metric
1 cup flour	=	4 oz	100 g
1 cup cocoa	=	4 oz	100 g
1 cup confectioner's sugar	=	4 oz	100 g
1 cup hard cheese – grated	=	4 oz	100 g
1 cup fruit – dried, e.g. raisins	=	4 oz	100 g
1 cup nuts – chopped	=	4 oz	100 g
1 cup biscuit crumbs	=	4 oz	100 g
1 cup coconut – grated	=	3 oz	75 g
1 cup breadcrumbs	=	3 oz	75 g
1 cup oats	=	3 oz	75 g

* Note that these are all approximate measurements. For more accurate results weigh the ingredients with scales.

FOOD GROUP 1:
Staples

Staple foods of a country are the main part of the diet, as they are cheap and usually easy to grow. They supply a large proportion of total energy for the body, from carbohydrates.

Staple foods include cereals, which are the seeds of certain grass plants, e.g. wheat, maize, oats and rice, which are milled to produce food products, e.g. flour, bread, pasta, puddings and breakfast cereals. Also included in this category are the starchy fruit and vegetables, e.g. plantain, breadfruit, sweet potato, cassava and yam.

The Caribbean diet has more than one staple food in general use, for instance both potato and corn bread are made and a lot of rice is eaten. As the foods from this group are rich in starch, the amount served will vary according to the individual. Staple foods need to be eaten with foods from the other groups to supply the missing nutrients.

Small amounts of protein, fat and vitamin B can be obtained from some of the staple foods, especially the cereals. Dietary fibre (roughage) is also provided by whole grain cereals, which have not had the outer covering or husk removed. These 'brown' equivalents, e.g. brown rice, wholewheat bread and bran cereals, are preferable to many refined foods, as they retain their natural fibre content. Recipes using high fibre foods are important for general good health and help to prevent constipation and other related diseases.

Wholewheat Bread

Imported wholewheat flour and dried yeast is skilfully kneaded and proved before baking to produce the fresh aroma and flavour of home-made bread. The addition of vitamin C enables a 'quick' bread to be produced.

Quantity			Ingredients	Equipment
1 lb	450 g	2 C	Wholewheat flour	Baking pans – flat
			or $\frac{1}{2}$ all-purpose flour	or 2 × 1 lb/900 g size

Quantity			Ingredients	Equipment
1 tsp	5 ml	1 tsp	Salt	Mixing bowl
1 oz	25 g	1 tblsp	Margarine	Teaspoon
1	1	1	Dried yeast – ¼ oz/ 7 g packet	Measuring jug
1 tsp	5 ml	1 tsp	Sugar	Wooden spoon Table knife
½ pt	250 ml	1 C	Water – warm	Flour dredger Plastic wrap – to cover dough
1	1	1	Vitamin C tablet (ascorbic acid)	Cooling rack

Variation For a sweet bread use all-purpose flour, add more sugar, some raisins and mix together with milk instead of water.

Method
1. Grease the baking pans.
2. Place the flour into a mixing bowl. (Sieve if using all-purpose flour.) Add the salt and rub in the margarine using the fingertips.
3. Add the yeast with the sugar and crushed vitamin C tablet to the warm water. If possible set aside for 5 minutes to become frothy.
4. Mix all the ingredients together and knead on a floured surface for 10 minutes.
5. Shape into 12 rolls or a loaf and cover. Leave in a warm place to prove, until the dough doubles in size. Set the oven to 425°F.
6. Remove the covering and bake for 20 minutes (small rolls) or up to 40 minutes (large loaf) depending on size. Reduce the oven to 375°F when the bread is well risen to complete cooking. The bread should sound 'hollow' when lightly tapped underneath. Place on a cooling rack straight away.

Serve with home-made *Lemon Cheese* or with a wedge of hard cheese and salad.

* Made by the traditional method the dough is 'knocked back' and left to prove one more time before baking.

Raisin Bread Pudding

A crisp bread topping, with a set egg custard mixture and fruit makes a dessert cake.

Quantity			Ingredients	Equipment
10	10	10	Wholewheat bread slices	Baking pan
				Foil wrap
2 oz	50 g	½ C	Raisins	Mixing bowl
2 oz	50 g	¼ C	Sugar	Wooden spoon
1	1	1	Egg	Measuring jug
1	1	1	Evaporated milk – 14 oz/400 g can size	Can-opener
1 tsp	5 ml	1 tsp	Vanilla flavouring	Teaspoon
½ tsp	2 ml	½ tsp	Cinnamon	Fork
4	4	4	Bananas – small size	
				Potato masher
1 oz	25 g	1 tblsp	Butter or margarine	Tablespoon

Variation Add canned fruit or the grated rind and juice of an orange instead of bananas. For a savoury bread pudding add 4 oz/100 g of grated cheese.

Method
1. Set the oven to 350°F and grease the baking pan.
2. Break up the bread into small pieces and place in a mixing bowl with raisins and sugar.
3. Beat the egg together with the milk, vanilla and cinnamon. Add to the bread.
4. Mash the bananas with the butter and add to the bread mixture. Mix well and place in the baking pan. Leave to soak if possible for 5–10 minutes.
5. Bake for 40 minutes and cover if the top begins to brown before the centre is cooked.

Serve cut into wedges as a cake.

Johnny Cake

Once called 'Journey' cake by the early settlers and eaten by fishermen of the islands, this soda bread is similar to scones or plain muffins.

Quantity			Ingredients	Equipment
8 oz	225 g	2 C	Flour	Baking pan – flat size

Quantity			Ingredients	Equipment
3 tsp	15 ml	3 tsp	Baking powder	Mixing bowl
small amount			Salt	Sieve
2 oz	50 g	$\frac{1}{4}$ C	Margarine or butter	Teaspoon Tablespoon
2 oz	50 g	$\frac{1}{4}$ C	Sugar	Measuring jug Wooden spoon
$\frac{1}{4}$ pt	125 ml	$\frac{1}{2}$ C	Milk or egg and milk	Table knife Rolling pin Flour dredger Biscuit cutter Cooling rack

Variation Add a little mixed spice or 2 oz/50 g of dried mixed fruit, grated coconut or cheese to the mixture.

Method
1. Set the oven to 400°F and grease the baking pan.
2. Sieve the flour and the dry ingredients into a mixing bowl.
3. Rub the margarine into the flour, using the fingertips. Add the sugar and any additional flavourings.
4. Slowly add the milk and mix to a soft dough consistency.
5. Shape into a round or make into small shapes, using a rolling pin and cutter. Place on the baking pan and brush with milk to glaze.
6. Bake for 15–25 minutes depending on the size. Remove from the oven and use a cooling rack.

Serve buttered with jelly or as an accompaniment to a savoury supper dish.

Corn Bread Muffins

Making use of a common staple food, maize grains can be popped (fried) or ground into flour to make products, such as cornflakes, corn-chips, cornstarch and cornmeal.

Quantity			Ingredients	Equipment
4 oz	100 g	1 C	Flour	Muffin or cup cake baking pan
5 oz	125 g	1 C	Cornmeal	Paper or foil cup cake cases
3 tsp	15 ml	3 tsp	Baking powder	Mixing bowl
2 oz	50 g	$\frac{1}{4}$ C	Sugar	Teaspoon
1	1	1	Egg	Measuring jug

Quantity			Ingredients	Equipment
6 fl oz	185 ml	$\frac{3}{4}$ C	Milk	Fork
2 oz	50 g	$\frac{1}{4}$ C	Cooking oil or margarine – melted	Wooden spoon or hand whisk
4 tblsp	60 ml	4 tblsp	Jelly/jam – fruit flavoured	Tablespoon

Variation Making chocolate flavoured muffins by replacing 1 oz/25 g of flour with the same amount of sifted cocoa powder.

Method
1. Set the oven to 400°F and line the baking pan with the cup cake cases.
2. Place the dry ingredients into a mixing bowl.
3. Mix together the egg, milk and cooking oil.
4. Slowly pour the milk mixture into the flour mixture and beat well.
5. Spoon half the batter into the muffin pans, until each pan is one-third full. Place a teaspoon of the jelly on to the batter and fill with the remaining batter.
6. Bake for 20–25 minutes until golden brown.

Serve warm for breakfast with butter and tea beverage.

Sweet Potato Pone

Sweet potatoes or cassavas are commonly grated and flavoured with cinnamon and spices to make a pudding or 'pone' as well as biscuits and savoury dishes.

Quantity			Ingredients	Equipment
3 oz	75 g	1 C	Coconut – grated	Baking dish – 9 × 12 ins/22$\frac{1}{2}$ × 30 cm size
1 lb	450 g	2 C	Sweet potato or cassava	Mixing bowl
4 oz	100 g	$\frac{1}{2}$ C	Sugar	Sharp knife Cutting board
2 oz	50 g	$\frac{1}{4}$ C	Butter or margarine	Cooking pot
2	2	2	Eggs	Potato masher
$\frac{1}{2}$ pt	250 ml	1 C	Milk	Wooden spoon
$\frac{1}{2}$ tsp	2 ml	$\frac{1}{2}$ tsp	Cinnamon	Measuring jug
$\frac{1}{2}$ tsp	2 ml	$\frac{1}{2}$ tsp	Mixed spice	Teaspoon

Variation Bake the pudding in a pastry case to make a pie.

Method
1. Set the oven to 375°F and grease the baking dish.
2. Soak the grated coconut in a little warm water to soften. Slice the sweet potatoes and boil in their skins for 20 minutes, until soft.
3. Remove the skins from the sweet potatoes and mash. Add the coconut and sugar.
4. Mix in the melted butter, eggs and milk.
5. Add the spice and beat the mixture until creamy. Pour the mixture into the baking dish and bake for 35 minutes until golden brown.

Serve hot or cold, dusted with confectioner's sugar or with an orange flavoured sauce.

Cheesy Staple Pie

A dietary staple of any type can be cooked and layered with a cheese sauce.

Quantity			Ingredients	Equipment
1 lb	450 g	2 C	Yam/cassava/breadfruit or eddoe/dasheen/tannia	Pie dish – 9 ins/ 23 cm size
1 oz	25 g	1 tblsp	Margarine	Cooking pots – 2
				Potato peeler
1 oz	25 g	2 tblsp	Flour	Sharp knife
¼ pt	125 ml	½ C	Milk	Cutting board
				Tablespoon
Small amount			Salt and pepper	Wooden spoon
				Measuring jug
1 tsp	5 ml	1 tsp	Mustard – prepared	Teaspoon
4 oz	100 g	½ C	Cheese – grated	Grater
1	1	1	Potato chips – packet, small size	Strainer
				Potato masher

Variation Add a little sautéd onion and sweet pepper.

Method
1. Wash and peel the staple food to be used. Cut into pieces and boil until tender for 20–30 minutes.
2. Set the oven to 350°F and grease the pie dish.
3. Prepare the sauce: Heat the margarine and add the flour. Cook for a few minutes on a low heat. Slowly add the milk off

the heat. Season with salt and pepper and the mustard, and continue stirring with a wooden spoon on low heat.
4. Grate the cheese and add two-thirds of this off the heat to the thickened sauce.
5. Strain the staple food and mash or cut into strips.
6. Layer the cooked staple food alternately with the cheese sauce in the pie dish, finishing with the sauce on top. Sprinkle with the crushed potato chips.
7. Bake for 20-25 minutes or brown under a hot broiler/grill.

Serve with sweet peas or green salad.

*Sauces can also be made by the all-in-one method: whisking together all the ingredients in a cooking pot until thick.

Savoury Macaroni

The many shapes and sizes of pasta combine well with a savoury tomato and ground beef mixture.

Quantity			Ingredients	Equipment
8 oz	225 g	2 C	Macaroni	Serving dish
1	1	1	Onion - small size	Cooking pot
1	1	1	Carrot	Sharp knife
1	1	1	Tomato	Cutting board
1 oz	25 g	1 tblsp	Margarine or cooking oil	Frying pan
8 oz	225 g	1 C	Ground beef/hamburger or corned beef - canned	Wooden spoon Can-opener
Small amount			Salt and pepper	Teaspoon
1 tsp	5 ml	1 tsp	Parsley - chopped	Measuring jug
2 tsp	10 ml	2 tsp	Soya sauce	Strainer

Variation Use a stock cube and tomato paste for extra flavour.

Method
1. Cook the macaroni in boiling salted water for 10 minutes.
2. Finely chop the onion and dice the carrot and tomato.
3. Sauté the vegetables in the hot margarine. Add the meat and allow it to brown.
4. Season and add a small amount of water. Simmer until the vegetables are tender.

5. Stir in the soya sauce and the cooked, drained macaroni and continue cooking for a further 5 minutes before serving.

Serve with warmed bread and salad.

Curried Plantain

Plantains are larger than a dessert banana and should be eaten cooked and usually served with meat.

Quantity			Ingredients	Equipment
6	6	6	Green bananas	Serving dish
1	1	1	Onion	Sharp knife
For frying			Margarine or cooking oil	Cutting board
2 tblsp	30 ml	2 tblsp	Curry powder	Cooking pot
½ oz	15 ml	1 tblsp	Flour	Tablespoon
½ pt	250 ml	1 C	Stock – vegetable or meat	Measuring jug
Small amount			Salt and pepper	

Variation Instead of plain boiled rice, serve with cooked brown rice mixed with 1 oz/25 g each of raisins and almonds.

Method
1. Peel and slice or grate the bananas.
2. Chop the onion and sauté in the margarine.
3. Add the curry powder (to individual taste) and the bananas. Brown lightly.
4. Blend the flour with a little of the stock.
5. Add all of the ingredients and simmer for 30 minutes.

Serve with rice and garnished with wedges of hard-boiled eggs and chopped parsley.

Stuffed Breadfruit

This large green dietary staple with its creamy flesh has a bread-like taste and, being edible, once cooked can be used like a potato.

Quantity			Ingredients	Equipment
1	1	1	Breadfruit	Baking dish
				Foil wrap

Food group 1: Staples 21

Quantity			Ingredients	Equipment
8 oz	225 g	1 C	Pork – minced or ground beef/ hamburger	Cooking pot
1	1	1	Onion – small size	Sharp knife
1	1	1	Tomato	Cutting board
Small amount			Cooking oil	Frying pan
Small amount			Salt and pepper	Teaspoon
$\frac{1}{4}$ tsp	1 ml	$\frac{1}{4}$ tsp	Chives or mixed herbs	Tablespoon

Variation   Include some wholewheat breadcrumbs in the stuffing mixture.

Method
1. Set the oven to 350°F and grease the baking pan.
2. Peel and par-boil the whole breadfruit in salted water.
3. Lightly fry the meat with the chopped vegetables in the cooking oil. Add the seasoning.
4. Cool the breadfruit and scoop out the core from the stalk end. Cut a small slice from the bottom, so that it stands in the baking pan.
5. Fill with the meat mixture, cover and bake for 45 minutes.

Serve hot with butter and garnished with chopped sweet pepper.

Jambalaya Rice

Rice makes a popular accompaniment to the main meal of the day as a simple rice salad or an 'all-in one' dish which is also called a *Paella, Pilaff, Pilau,* or *Risotto.*

Quantity			Ingredients	Equipment
8 oz	225 g	1 C	Ham – joint or slices	Serving dish
4 oz	100 g	$\frac{1}{2}$ C	Sausage or cooked chicken	Sharp knife
1	1	1	Onion	Cutting board
2	2	2	Garlic cloves	Cooking pot – large size with lid
1	1	1	Sweet pepper	Wooden spoon
2	2	2	Celery sticks	
4	4	4	Tomatoes – fresh	Measuring jug

Quantity			Ingredients	Equipment
1½ pts	750 ml	3 C	Beef or chicken stock water	Teaspoon
1 lb	450 g	2 C	Rice – uncooked	
Small amount			Salt and pepper	
½ tsp	2 ml	½ tsp	Cayenne pepper	

Variation This dish is especially good for using left-over foods and your imagination!

Method
1. Chop the meat and lightly fry in the cooking oil.
2. Finely chop the vegetables and add to the cooking pot to sauté.
3. Add the tomatoes with the stock or water and the rice.
4. Stir in the seasoning and bring to the boil. Cover and simmer for 15 minutes.
5. Stir, adding more water if required, re-cover and cook for 15 minutes longer.

Serve as a meal in itself with freshly cooked green vegetables.

* When cooking rice to accompany meat or fish, to every cup measure of rice add 2 cups of water.

FOOD GROUP 2:
Legumes

Legumes or pulses are the dried seeds of leguminous plants, e.g. pigeon (gungo) peas, black-eye peas, split peas, soya beans, kidney beans and also peanuts (ground nuts). There is a colourful, wide range available.

Legumes are sometimes placed in a group labelled 'meat and alternatives', as they are a very good source of protein for body building. We do not usually use enough peas and beans in our meals although they are a low-cost nutritious food which can be easily grown. Their protein value is made complete when eaten with cereals, for example peas and rice, or baked beans (haricot beans, cooked and canned in tomato sauce) with toasted bread.

Soya beans contain good quality protein and are used to manufacture textured vegetable protein (TVP), which can look and taste like meat. This is used as a meat substitute in the making of economical meat products, e.g. hamburgers, sausages and meat loaf. A strict vegetarian or vegan will depend on legumes and other vegetable protein sources, including cereal grains, in the diet. Legumes also contain carbohydrate and some fibre, vitamins and minerals.

Dried pulses need more preparation and careful cooking than the easy to use canned peas and beans available. They need soaking overnight and long cooking – for up to two hours. Avoid adding salt during the cooking process as this will toughen the skins. If possible use a pressure cooker to shorten the cooking time.

In this section there are some excellent recipe ideas for vegetarians and cheaper protein dishes to make.

Vegetable Hot-Pot

A colourful collection of fresh vegetables, peas and beans are made into a stew and topped with herb dumplings.

Quantity			Ingredients	Equipment
1	1	1	Onion	Casserole dish

Quantity			Ingredients	Equipment
1	1	1	Garlic clove	Sharp knife
1	1	1	Potato – large size	Cutting board
2	2	2	Carrots	Frying pan
1 lb	450 g	2 C	Pumpkin	Wooden spoon
4 oz	100 g	$\frac{1}{2}$ C	Butter beans – canned or dried – soaked and cooked	Can-opener
4 oz	100 g	$\frac{1}{2}$ C	Peas – canned or fresh shelled	Tablespoon
For frying			Cooking oil	Measuring jug
1 oz	25 g	2 tblsp	Flour	Teaspoon
$1\frac{1}{2}$ pts	750 ml	3 C	Vegetable stock	Mixing bowl
2 tblsp	30 ml	2 tblsp	Tomato paste	Table knife
$\frac{1}{2}$ tsp	2 ml	$\frac{1}{2}$ tsp	Thyme or mixed herbs	
Small amount			Salt and pepper	

For dumplings:

8 oz	225 g	2 C	Wholewheat flour	
1 tsp	5 ml	1 tsp	Baking powder	
$\frac{1}{2}$ tsp	2 ml	$\frac{1}{2}$ tsp	Mixed herbs	
1 tblsp	15 ml	1 tblsp	Cooking oil	
$\frac{1}{4}$ pt	125 ml	$\frac{1}{2}$ C	Water	
1 tblsp	15 ml	1 tblsp	Skimmed milk – optional	

Variation   Add curry powder to make a hot flavoured gravy and a teaspoon of molasses to give a darker colour.

Method
1. Set the oven to 350°F and grease the casserole dish.
2. Wash, peel and chop all the vegetables, removing the inedible outer parts and seeds.
3. Sauté the chopped onion and garlic in the cooking oil. Place in the dish with the other vegetables.
4. Add a little more cooking oil to the frying pan if necessary and stir in the flour. Add the stock, tomato paste and seasoning and simmer for a few minutes.
5. Pour the gravy over the vegetables and bake for 45 minutes.

6. To make the dumplings: Combine the flour, baking powder and mixed herbs. Mix in the oil and water to make a soft dough. Dust the board with flour and make the dough into 8 small rounds. Add to the casserole dish and bake for a further 15-20 minutes.

Serve sprinkled with chopped parsley.

Three Bean Salad

Using canned or soaked (overnight) and cooked dried beans, this is a successful salad in its own French dressing.

Quantity			Ingredients	Equipment
1	1	1	Green beans – 16 oz/450 g can size	Serving bowl and cover
1	1	1	Wax beans – 16 oz/450 g can size	Can-opener
1	1	1	Red kidney beans – 16 oz/450 g can size	Mixing bowl
1	1	1	Onion	Tablespoon
1	1	1	Garlic clove	Sharp knife
$\frac{1}{2}$	$\frac{1}{2}$	$\frac{1}{2}$	Sweet pepper	Cutting board

For French dressing:

2 tblsp	30 ml	2 tblsp	Lime juice	Bottle or container with lid
4 tblsp	60 ml	4 tblsp	Salad oil	
1 oz	25 g	1 tblsp	Sugar	
Small amount			Salt and pepper	
$\frac{1}{2}$ tsp	2 ml	$\frac{1}{2}$ tsp	Dry mustard	

Variation Use the French dressing for a tossed green salad: lettuce, cucumber, sweet pepper, etc.

Method
1. Mix together the drained beans in a mixing bowl.
2. Slice the onion and finely chop the garlic and sweet pepper.
3. Combine the dressing ingredients by shaking together in a

covered container. Pour over the salad and marinate overnight in the refrigerator.

Serve as an appetiser or with the main meal. Store in a sealed container in the refrigerator.

Chilli Sauce

A good chilli sauce should be moist on completion, not too 'runny', and well flavoured. Keep tasting to discover how hot and spicy you are making it!

Quantity			Ingredients	Equipment
1	1	1	Onion – small size	Container with cover
1	1	1	Sweet pepper	Sharp knife
For frying			Cooking oil	
1	1	1	Red kidney beans – 16 oz/450 g can size	Cutting board
1	1	1	Tomatoes – 14 oz/ 400 g can size	Cooking pot
2 tblsp	30 ml	2 tblsp	Tomato paste	Wooden spoon
1 tblsp	15 ml	1 tblsp	Chilli powder or small hot peppers	Can-opener
1	1	1	Bay leaf – crushed	Tablespoon
Small amount			Salt and pepper	
$\frac{1}{2}$ tblsp	7 ml	$\frac{1}{2}$ tblsp	Brown sugar	Teaspoon
To taste			Hot pepper sauce	

Variation Add 1 lb/450 g hamburger/ground beef or chopped stewing steak to the cooking pot. Use a can of baked beans in tomato sauce for a quick chilli sauce.

Method
1. Dice the onion and sweet pepper and sauté in the cooking oil. (Add the meat, if you are using any, at this stage.)
2. Add the cooked beans, canned tomatoes, tomato paste, seasoning and brown sugar.
3. Cover the cooking pot and simmer for 30 minutes. Add hot pepper sauce to taste if necessary.

Serve with warm corn bread or garlic flavoured bread.

Pulse Burgers

Always popular, burgers or meat patties are a quick meal, served on a soft bread roll with cheese, salad and tomato relish.

Quantity			Ingredients	Equipment
1	1	1	Red kidney beans – 16 oz/450 g can size	Serving dish Can-opener
1	1	1	Onion – small size	Mixing bowl
8 oz	225 g	1 C	Brown rice – cooked	Potato masher
$\frac{1}{2}$ tsp	2 ml	$\frac{1}{2}$ tsp	Chilli powder	Sharp knife
Small amount			Salt and pepper	Cutting board
1 tsp	5 ml	1 tsp	Soya sauce	Teaspoon
Small amount			Flour	Tablespoon
				Flour dredger
				Fork
1	1	1	Egg	Plates – 2
3 oz	75 g	1 C	Oats or whole-wheat breadcrumbs	Frying pan
For frying			Cooking oil	Draining spoon Kitchen paper

Variation Instead of kidney beans use washed brown or green lentils, which have been cooked in vegetable stock for 25 minutes (or until all of the liquid is absorbed), and add 1 teaspoon of tomato paste to flavour.

Method
1. Drain the beans, place in a bowl and mash.
2. Finely chop the onion and add to the beans with the cooked rice and seasoning.
3. Shape the mixture into 6 ball shapes, then flatten into patties. Use a little flour if necessary to prevent sticking.
4. Dip the patties into beaten egg and then into the oats.
5. Heat a small amount of cooking oil in a frying pan and shallow fry the patties for 4 minutes on each side or place under a hot broiler/grill. Drain on kitchen paper after frying.

Serve garnished with crisp lettuce or watercress. Freeze raw patties between layers of waxed paper.

Basic Bean Bake

Home-made baked beans take a long time to bake so try this high quality protein dish instead. If using dried beans, soak them overnight, avoid adding salt, and cook for at least two hours – or pressure cook in less time.

Quantity			Ingredients	Equipment
1	1	1	Beans – 16 oz/ 450 g can size e.g. soya beans, white haricot, butter beans	Baking dish
1	1	1	Sweet-corn – 16 oz /450 g can size	Can-opener
4	4	4	Tomatoes – fresh or canned	Mixing bowl
1	1	1	Onion – small size	Tablespoon Bowl – small size
1 oz	25 g	1 tblsp	Butter or margarine	Sharp knife
Small amount			Salt and pepper	Cutting board
4	4	4	Bacon rashers	Teaspoon
3 tblsp	45 ml	3 tblsp	Mayonnaise	Whisk
2 oz	50 g	$\frac{2}{3}$ C	Wholewheat breadcrumbs	
1 tsp	5 ml	1 tsp	Soya sauce or hot sauce	
$\frac{1}{2}$ tsp	2 ml	$\frac{1}{2}$ tsp	Cayenne pepper	
1	1	1	Egg white	
$\frac{1}{2}$ tsp	2 ml	$\frac{1}{2}$ tsp	Parsley or mixed herbs	

Variation Add chopped apple instead of tomatoes and use grated cheese for the topping instead of the egg mixture.

Method
1. Set the oven to 350°F and grease the baking dish.
2. Drain the canned beans and sweet-corn and mix together. Place in the baking dish.
3. Skin the tomatoes by placing in boiling water for about a minute depending on how ripe they are, then slice them. Place in dish over the beans and sweet-corn.
4. Chop the onion and place over the top. Dot with butter and sprinkle with salt and pepper.

5. Arrange the bacon over the onion and bake for 20 minutes.
6. Blend the remaining ingredients and whisk the egg white until stiff. Fold into the mayonnaise mixture and use to cover the bacon.
7. Sprinkle with the chopped parsley and return to the oven for 10–15 minutes, until the egg mixture is firm but still pale in colour.

Serve with toasted bread and a green salad.

Pigeon Peas and Rice

Soft brown or the familiar red 'peas' in Jamaica combine well with the popular starchy accompaniment to the main meal of the day.

Quantity			Ingredients	Equipment
4 oz	100 g	$\frac{1}{4}$ C	Salt pork	Serving bowl with cover
1 tblsp	15 ml	1 tblsp	Cooking oil	Cooking pot – large size
$\frac{1}{2}$	$\frac{1}{2}$	$\frac{1}{2}$	Onion – small size	Sharp knife
$\frac{1}{4}$	$\frac{1}{4}$	$\frac{1}{4}$	Sweet pepper	Cutting board
$\frac{1}{2}$	$\frac{1}{2}$	$\frac{1}{2}$	Celery stick	Tablespoon
2 tblsp	30 ml	2 tblsp	Tomato paste	Teaspoon
1 tsp	5 ml	1 tsp	Thyme – chopped	Can-opener
Small amount			Salt and pepper	Measuring jug
1	1	1	Pigeon peas/gungo peas – 10 oz/275 g can size or red kidney beans	
1 pt	500 ml	2 C	Water	
8 oz	225 g	1 C	Rice – uncooked	

Variation Cook the rice in coconut milk instead of water and flavour with a little curry powder.

Method
1. Fry the salt pork in a cooking pot and add the oil.
2. Finely chop the onion, sweet pepper and celery.
3. Sauté the vegetables and add the tomato paste and thyme. Allow to simmer for 5 minutes.

4. Add the drained pigeon peas and pour in the water. Season to taste.
5. Bring to the boil and add the rice. Continue cooking over a medium heat for 20–30 minutes until the rice is tender. Add a little more water if necessary and remove any remaining salt pork before serving.

Serve as a base for curry, meat or fish dishes.

Carrot Pea Soup

A filling soup made from carrots, containing carotene (vitamin A) and peas which are a rich source of vegetable protein.

Quantity			Ingredients	Equipment
2	2	2	Carrots	Flask or container with cover
1	1	1	Onion	Sharp knife
4 oz	100 g	$\frac{1}{2}$ C	Salt pork or streaky bacon	Cutting board
1 lb	450 g	2 C	Split peas – soaked	Cooking pot – large size with lid
2 pts	1 lit	4 C	Water	Measuring jug
$\frac{1}{4}$ pt	125 ml	1 C	Milk	Tablespoon
$\frac{1}{4}$ tsp	1 ml	$\frac{1}{4}$ tsp	Allspice	Teaspoon
Small amount			Black pepper	Sieve or electric blender
1	1	1	Bay leaf	
1 tsp	5 ml	1 tsp	Soya sauce	
Small amount			Celery salt	
Small amount			Garlic salt	

Variation Use a ham bone instead of salt pork and add a little peanut butter for flavour.

Method
1. Finely chop the carrots and onion.
2. Heat the salt pork in a cooking pot and add the chopped vegetables, peas which have been soaked overnight and the water. Season.
3. Cover the cooking pot and simmer for 45 minutes or until the peas are tender.
4. Discard any remaining salt pork. (Remove any meat from the

ham bone if using.) Add the milk and check the seasoning to taste. Sieve or blend to make a smooth soup if liked.

Serve garnished with *croûtons* (fried cubes of bread) as an appetiser or for lunch.

Nut Roast

Use unsalted shelled nuts and seeds to produce a crumbly moist loaf served with a vegetarian gravy sauce, as an alternative to the familiar joint of meat.

Quantity			Ingredients	Equipment
12 oz	350 g	$1\frac{1}{2}$ C	Mixed nuts and seeds e.g. cashew nuts, almonds, hazelnuts, pecans, pumpkin seeds, sesame seeds	Loaf pan – 2 lb/ 900 g size Foil wrap Rolling pin Plastic bag
1	1	1	Onion	Sharp knife
1	1	1	Celery stick	Cutting board
6 oz	150 g	2 C	Wholewheat breadcrumbs	Tablespoon
1 tblsp	15 ml	1 tblsp	Soya sauce	Teaspoon
1 tsp	5 ml	1 tsp	Mixed herbs	
Small amount			Salt and pepper	
2	2	2	Eggs	Cooking pot
5 tblsp	75 ml	5 tblsp	Cooking oil	Measuring jug
Extra to garnish			Pumpkin seeds	
For the gravy sauce:				
1	1	1	Onion – small size	
1 oz	25 g	1 tblsp	Cooking oil or margarine	
1 oz	25 g	2 tblsp	Flour	
$\frac{3}{4}$ pt	350 ml	$1\frac{1}{2}$ C	Vegetable stock	
1 tsp	5 ml	1 tsp	Soya sauce	
1 tsp	5 ml	1 tsp	Yeast extract	
1 tsp	5 ml	1 tsp	Tomato paste	

Variation Add chopped mushrooms to the nut loaf mixture.

Method
1. Set the oven to 375°F and grease the loaf pan.
2. Crush or finely chop the nuts and seeds with the onion and celery. * Use a plastic bag and rolling pin to crush the nuts.
3. Mix all the ingredients together, squeezing with the hands. Reserve 2 tablespoons of the cooking oil for the top.
4. Place the mixture into the loaf pan. Pour the reserved oil over the top and sprinkle with additional pumpkin seeds.
5. Cover with foil wrap and bake for 45 minutes.
6. For the gravy finely chop the onion. Heat the cooking oil with the onion and flour until beginning to turn brown. Slowly add the stock off the heat. Add the seasoning and simmer for 5 minutes.

Serve the nut roast with this brown sauce and baked potatoes, carrots and a green vegetable.

Lentil Curry Spread

Versatile lentils make excellent vegetarian bakes, soups and dips, having a 'nutty' taste and being of good protein value, without the same preparation and cooking time as dried beans.

Quantity			Ingredients	Equipment
8 oz	225 g	1 C	Lentils – brown or green	Serving bowl – small size
1 pt	500 ml	2 C	Vegetable stock	Cooking pot
1	1	1	Onion	Measuring jug
1	1	1	Garlic clove	Sharp knife
1 oz	25 g	1 tblsp	Margarine or butter	Cutting board
Small amount			Salt and pepper	Tablespoon
				Wooden spoon
1 tblsp	15 ml	1 tblsp	Curry powder	Fruit squeezer
1	1	1	Lime	Teaspoon
1 tsp	5 ml	1 tsp	Parsley	

Variation Add 1 tablespoon of grated coconut or tomato paste for a different flavour.

Method
1. Cook the lentils in the stock for 25 minutes (or until all of the liquid is absorbed).

2. Finely chop the onion and garlic and sauté in the hot margarine. Add the seasoning, stir in the curry powder and cook for a few minutes.
3. Squeeze the juice from half of the lime. Mix this in with the cooked lentils and chopped parsley, off the heat.
4. Place the mixture into the serving bowl and garnish with twists of sliced lime.

Serve over rice or in small pots for spreading on toasted bread.

Peanut Butter Cookies

Delicious small biscuits which melt in the mouth – be sure to make two batches!

Quantity			Ingredients	Equipment
4 oz	100 g	$\frac{1}{2}$ C	Peanut butter	Container with cover
4 oz	100 g	$\frac{1}{2}$ C	Butter or margarine	Baking pans – 2 flat size
6 oz	150 g	$\frac{2}{3}$ C	Brown sugar	
1	1	1	Egg	Mixing bowl
$\frac{1}{2}$ tsp	2 ml	$\frac{1}{2}$ tsp	Vanilla flavouring	Wooden spoon
5 oz	125 g	$\frac{2}{3}$ C	Flour	Teaspoon
Small amount			Salt	Sieve
$\frac{1}{2}$ tsp	2 ml	$\frac{1}{2}$ tsp	Baking powder	Fork

Variation Instead of peanut butter flavour make coconut, raisin or chocolate chip cookies.

Method
1. Set the oven to 375°F and grease the baking pans.
2. In a mixing bowl cream together the butter and sugar.
3. Add the egg, vanilla and dry ingredients and mix well. Form into a dough using the hands. Allow to chill in the refrigerator if possible.
4. Form into small ball shapes and place on the baking pans. Flatten with a fork dipped in flour.
5. Bake for 10–15 minutes. Allow to harden on the baking pan before removing.

Serve as an appetiser or with fruit punch for dessert.

FOOD GROUP 3:
Dark green leafy/ yellow vegetables

Dark green leafy and other non-starchy vegetables include: cabbage, callaloo, dasheen leaves, broccoli, pumpkin, squash, carrot and tomato. They are an important source of vitamins and minerals, especially vitamins A and C, in our meals. These micronutrients are only required in small amounts, but are still essential to keep the body healthy.

Good storage and cooking of fresh vegetables is essential. Green vegetables deteriorate very quickly and should be stored in the refrigerator in a plastic bag. Wash thoroughly before use to remove dirt and insects. To conserve the food value, cook the vegetables in a small amount of water for a short period of time. It is sometimes a good idea to cut vegetables into small pieces in order to cook them quickly and retain maximum food value. Serve at once and avoid reheating. If possible use the cooking water for vegetable stock, which is useful in making sauces and soups.

Vegetables add variety in colour, texture and flavour to the diet. Serve raw vegetable and fruit salad mixtures to provide a better supply of fibre. Also add herbs, which are green plants with a distinctive taste or smell and are used to flavour savoury foods. For example thyme is a frequently used herb in these Caribbean recipes.

Broccoli and Cheese Sauce

Broccoli, calabrese and cauliflower are the flowering shoots from plants belonging to the cabbage family, and are also served as a vegetable with melted butter and lime juice or stir-fried with almonds.

Quantity			Ingredients	Equipment
3	3	3	Broccoli heads or	Baking dish
1	1	1	Cauliflower	Sharp knife
				Cutting board
1 oz	25 g	1 tblsp	Margarine or butter	Cooking pots – 2

Food group 3: Dark green leafy/yellow vegetables

Quantity			Ingredients	Equipment
1 oz	25 g	2 tblsp	Flour	Tablespoon
½ pt	250 ml	1 C	Milk	Wooden spoon
Small amount			Salt and pepper	Measuring jug
4 oz	100 g	½ C	Cheese – grated	Grater
2 tblsp	30 ml	2 tblsp	Wholewheat breadcrumbs	Strainer

Variation Use a packet of frozen mixed vegetables to coat with the cheese sauce.

Method
1. Wash the broccoli and trim the stalks. (Remove any outer leaves from the cauliflower and leave it whole or cut into florets.)
2. Half fill a cooking pot with water, add the broccoli and bring to boil. Simmer for 15 minutes until tender.
3. Make the sauce by placing the margarine and flour in a cooking pot on a low heat to make a 'roux'. Cook for a few minutes and remove from the heat. Slowly add the milk. Season and return to the heat, stirring until the sauce thickens.
4. Grate the cheese and add two-thirds of this off the heat to the sauce. Place the strained broccoli into a greased baking dish.
5. Pour the cheese sauce over the broccoli and sprinkle with the remaining cheese and breadcrumbs. Brown under a hot broiler/grill or in the oven.

Serve as a supper dish with crisp French bread or as a vegetable accompaniment to a meat or fish dish.

Egg-Plant Bake

Also called melongene or aubergine, this smooth purple-skinned vegetable cooked with sweet peppers makes a good vegetarian dish or *Ratatouille*.

Quantity			Ingredients	Equipment
1	1	1	Egg-plant	Baking dish
4	4	4	Zucchini	Sharp knife
1	1	1	Onion – large size	Cutting board
1	1	1	Sweet pepper – large size	Frying pan – large size
1 oz	25 g	1 tblsp	Butter or margarine	Draining spoon

Quantity			Ingredients	Equipment
1–2 tblsp	15–30 ml	1–2 tblsp	Cooking oil	Kitchen paper
3	3	3	Tomatoes – fresh or large can size	Wooden spoon Can-opener
1	1	1		
2 tblsp	30 ml	2 tblsp	Tomato paste	Tablespoon
1 tsp	5 ml	1 tsp	Basil or mixed herbs	Teaspoon
Small amount			Salt and pepper	Grater
2 tblsp	30 ml	2 tblsp	Breadcrumbs	
2 oz	50 g	$\frac{1}{4}$ C	Cheese – grated	

Variation Cover with slices of hard cheese and top with bacon slices before baking.

Method
1. Set the oven to 375°F and grease the baking dish.
2. Wash and slice the egg-plant. Cover with salt and leave covered for 20 minutes (to extract some of the moisture).
3. Scrub the zucchini and slice with the other vegetables.
4. Fry the egg-plant in the butter and oil, until lightly brown. Remove from the pan and sauté the sliced vegetables.
5. Add the tomatoes, with a small amount of water if using fresh tomatoes, and the seasoning. Return the egg-plant to the pan and continue cooking for 5 minutes.
6. Place in the baking dish and cover with breadcrumbs mixed with grated cheese and bake for 30 minutes.

Serve as a supper dish. As this dish makes a large quantity it freezes very well.

Cooked Christophene

This delicately flavoured vegetable, similar to a squash, is pear-shaped and creamy in colour. An unripe papaya can also be used in this recipe and filled with a tomato and cheese mixture.

Quantity			Ingredients	Equipment
1	1	1	Christophene, papaya or pumpkin	Baking pan Foil wrap
1	1	1	Onion – small size	Sharp knife
2	2	2	Garlic cloves	Cutting board
$\frac{1}{2}$	$\frac{1}{2}$	$\frac{1}{2}$	Sweet pepper	Cooking pot
2	2	2	Tomatoes – fresh	Frying pan

Quantity			Ingredients	Equipment
For frying			Cooking oil	Wooden spoon
2	2	2	Ham or bacon slices	Scissors
				Mixing bowl
				Potato masher
1½ oz	40 g	½ C	Wholewheat breadcrumbs	Tablespoon
1	1	1	Egg	Teaspoon
Small amount			Salt and pepper	
¼ tsp	1 ml	¼ tsp	Nutmeg – grated	
Small amount			Water	

Variation Fill with a savoury meat mixture for '*Meat in the Moon*'.

Method
1. Set the oven to 350°F and grease the baking pan.
2. Wash the christophene and cut in half lengthwise. Par-boil for 20 minutes in salted water.
3. Chop the vegetables and sauté in the cooking oil. Remove from the pan and lightly fry the chopped bacon.
   * Use scissors to chop the meat more easily.
4. Scoop out the pulp from the cooked christophene and mash. Add the other ingredients and seasoning.
5. Spoon this mixture back into the hollowed-out vegetable shell. Place in the baking pan, with a little water to prevent sticking. Cover with foil wrap and bake for 30 minutes.

Serve with a sharp Cheddar cheese sauce or sour cream.

Cabbage Casserole

Large green cabbage leaves are cooked, filled with a tasty meat filling and served with a crushed tomato and herb sauce.

Quantity			Ingredients	Equipment
1	1	1	Cabbage-head – large size	Baking dish
				Foil wrap
1	1	1	Onion – small size	Strainer
½	½	½	Sweet pepper	Cooking pot
				Sharp knife
				Cutting board
8 oz	225 g	1 C	Ground beef/ hamburger	Mixing bowl
8 oz	225 g	1 C	Rice – cooked	Wooden spoon

Quantity			Ingredients	Equipment
1	1	1	Egg	Teaspoon
Small amount			Salt and pepper	Measuring jug
1 tsp	5 ml	1 tsp	Soya sauce	Cocktail sticks or tooth picks
$\frac{1}{2}$ pt	250 ml	1 C	Beef stock (stock cube and water)	Frying pan
1 oz	25 g	1 tblsp	Margarine	Tablespoon
1 oz	25 g	1 tblsp	Brown sugar	Can-opener
1 oz	25 g	2 tblsp	Flour	
1	1	1	Tomatoes – 14 oz/ 400 g can size or fresh	
$\frac{1}{2}$ tsp	2 ml	$\frac{1}{2}$ tsp	Basil or thyme – chopped	

Variation Use this mixture to stuff an egg-plant, and serve with a canned tomato sauce.

Method
1. Set the oven to 350°F and grease the baking dish.
2. Separate 8 outer leaves from the cabbage and place in boiling water. Leave for 2 minutes then drain.
3. Make the filling by finely chopping the onion and sweet pepper. Mix with the ground beef, rice, egg and seasoning in a bowl. Add a little of the beef stock if the mixture is too dry.
4. Use to fill each cabbage leaf. Fold in the two sides and roll up, securing with a cocktail stick if necessary.
5. Melt the margarine and sugar in a frying pan and brown the rolls on each side. Arrange inside the baking dish.
6. Blend the flour with the beef stock and add to the frying pan with the canned tomatoes and seasoning. Add a small amount of water if using fresh tomatoes. Bring to the boil, stirring until thickened.
7. Pour over the cabbage rolls. Bake covered for 1 hour or until the cabbage is tender.

Serve with sweet-corn for a contrast in colour.

Callaloo

The leafy spinach-like 'greens' from the dasheen root are called taro, and make a tasty soup with the addition of crab meat.

Food group 3: Dark green leafy/yellow vegetables

Quantity			Ingredients	Equipment
8 oz	225 g	1 C	Callaloo or spinach	Container with cover
1	1	1	Onion	Strainer
1	1	1	Garlic clove	Sharp knife
1 oz	25 g	1 tblsp	Butter or margarine	Cutting board
1 pt	500 ml	2 C	Vegetable or chicken stock	Wooden spoon Tablespoon
$\frac{1}{4}$ pt	125 ml	$\frac{1}{2}$ C	Coconut milk	Measuring jug
Small amount			Salt and pepper	Can-opener
8 oz	225 g	1 C	Crab meat – fresh or canned	
To taste			Hot sauce	

Variation An alternative use of callaloo or spinach is to cook it as a vegetable, instead of making into a soup.

Method
1. Wash the callaloo leaves discarding any damaged parts. Finely chop the onion and garlic.
2. Gently fry the onion and garlic until soft and add the callaloo to coat with the fat.
3. Stir in the stock, coconut milk and seasoning. Bring to the boil and simmer for 10 minutes until the vegetables are tender.
4. Stir in the crab meat and hot sauce to taste. Continue cooking to heat through.

Serve with freshly made wholewheat bread.

Cream of Pumpkin Soup

Pumpkin or calabaza contains mainly water, but is a good thickening agent for a soup base and a rich source of carotene (vitamin A).

Quantity			Ingredients	Equipment
1 lb	450 g	2 C	Pumpkin or calabaza	Flask or container with cover
1	1	1	Onion	Sharp knife
1	1	1	Celery stick	Cutting board
1 oz	25 g	1 tblsp	Margarine or butter	Cooking pot – large size

Quantity			Ingredients	Equipment
				Strainer
				Bowl-small size
1 oz	25 g	2 tblsp	Flour	Potato masher
$\frac{1}{2}$ pt	250 ml	1 C	Milk	Tablespoon
1 pt	500 ml	3 C	Vegetable or chicken stock	Wooden spoon
1	1	1	Bay leaf	Teaspoon
$\frac{1}{2}$ tsp	2 ml	$\frac{1}{2}$ tsp	Thyme – chopped	Measuring jug
Small amount			Salt and pepper	

Variation Use potatoes instead of pumpkin.

Method
1. Remove the peel and seeds from the pumpkin. Chop and cook for 20 minutes until tender. Strain and mash to a purée.
2. Sauté the onion and celery in the fat until soft. Stir in the flour.
3. Slowly add the milk off the heat, stirring all the time.
4. Return to the heat and add the pumpkin purée, milk, stock and seasoning to taste. Continue cooking to heat through.

Serve in individual bowls, garnished with chopped parsley.
*Use a pair of scissors to chop the parsley in a small bowl or cup.

Red Cabbage Slaw

Raw cabbage contains a lot of vitamin C and the red type gives added colour to a well liked salad.

Quantity			Ingredients	Equipment
$\frac{1}{2}$	$\frac{1}{2}$	$\frac{1}{2}$	Cabbage – red colour	Container with cover
2	2	2	Carrots	Strainer
1	1	1	Celery stick	Sharp knife
$\frac{1}{2}$	$\frac{1}{2}$	$\frac{1}{2}$	Onion – small size	Cutting board
1	1	1	Apple – red-skinned	Potato peeler
$\frac{1}{2}$	$\frac{1}{2}$	$\frac{1}{2}$	Lime	Grater
2 tblsp	30 ml	2 tblsp	Mayonnaise	Fruit squeezer
2 tblsp	30 ml	2 tblsp	Natural yogurt	Tablespoon
1 oz	25 g	$\frac{1}{4}$ C	Raisins	
Small amount			Salt and pepper	

Variation Add mandarin orange segments, sliced grapefruit or pineapple cubes to the salad.

Method
1. Wash the cabbage and cut into fine shreds or grate.
2. Peel, wash and grate the carrots. Finely chop or grate the other vegetables.
3. Add the lime juice to the chopped apple to prevent browning. Add enough mayonnaise and natural yogurt to bind the mixture together.
4. Add the raisins and seasoning and toss ingredients together. Chill in the refrigerator.

Serve as a side salad for meat and fish dishes. Garnish with chopped pecan nuts.

Sunshine Salad

A very appetising salad which with its protein-rich dressing makes a complete meal served with crunchy digestive biscuits.

Quantity			Ingredients	Equipment
2 oz	50 g	$\frac{1}{2}$ C	Raisins	Serving dish
2	2	2	Carrots – large size	Mixing bowl
4	4	4	Bananas – small size	Potato peeler
$\frac{1}{2}$	$\frac{1}{2}$	$\frac{1}{2}$	Lime	Grater
1	1	1	Watercress – bunch	Sharp knife
2	2	2	Oranges – large size	Cutting board
				Tablespoon
Coconut cream dressing:				
8 oz	225 g	1 C	Cottage cheese – carton	Fruit squeezer
1 oz	25 g	$\frac{1}{3}$ C	Coconut – grated	Strainer
$\frac{1}{4}$ pt	125 ml	$\frac{1}{2}$ C	Evaporated milk or single cream	Sieve
$\frac{1}{2}$ tsp	2 ml	$\frac{1}{2}$ tsp	Honey	Wooden spoon
				Teaspoon

Variation Serve on a base of crisp lettuce or chicory leaves instead of watercress.

Method

1. Wash the raisins and leave to soak in a small amount of warm water for 10 minutes.
2. Peel the carrots and grate. Skin and slice the bananas and toss in the lime juice with the carrots and raisins.
3. Place in the centre of the serving dish and surround with the washed watercress and sliced oranges (keeping the outer peel on the oranges).
4. For the dressing, sieve the cottage cheese and blend with the other ingredients in a bowl.

Serve with the coconut cream dressing over the top and garnished with toasted coconut or finely grated orange rind.

Okra and Tomatoes

These 'ladies fingers' also make a good rice soup, called *Gumbo*. Avoid any sliminess when cooking by keeping whole and adding a small amount of acid, such as vinegar or lime juice.

Quantity			Ingredients	Equipment
1 lb	450 g	2 C	Okras	Serving dish
1	1	1	Onion	Cooking pot
1	1	1	Garlic clove	Sharp knife
2	2	2	Tomatoes – large size	Cutting board
1 oz	25 g	1 tblsp	Margarine or butter	Wooden spoon
$\frac{1}{2}$	$\frac{1}{2}$	$\frac{1}{2}$	Lime	Fruit squeezer
$\frac{1}{2}$ tsp	2 ml	$\frac{1}{2}$ tsp	Hot sauce	Tablespoon
1–2	1–2	1–2	or hot peppers	Teaspoon
Small amount			Salt and pepper	

Variation Add a little chopped sweet pepper.

Method

1. Wash the okra and cut off the stalk end, but not the pod. Slice the onion and chop the other vegetables.
2. Sauté the okra and onion rings in the melted fat and lime juice. Stir until the sliminess disappears from the okra.
3. Add the tomatoes and seasoning and cook for 15 minutes, or until the vegetables are tender and the mixture thickens.

Serve as a vegetable accompaniment or as a lunch or supper dish, with grated cheese.

Frosted Pumpkin Cake

Another use for pumpkin is for cookies, pies, breads and a very moist cake, which is spread with a smooth cream-cheese frosting.

Quantity			Ingredients	Equipment
4	4	4	Eggs	Tube pan – 10 ins/ 25 cm size
1 lb	450 g	2 C	Sugar	Mixing bowl
$\frac{1}{2}$ pt	250 ml	1 C	Vegetable oil	Wooden spoon
8 oz	225 g	2 C	Flour	Measuring jug
2 tsp	10 ml	2 tsp	Baking soda	Sieve
2 tsp	10 ml	2 tsp	Cinnamon	Teaspoon
1 lb	450 g	2 C	Pumpkin – puréed fresh or canned	Can-opener Spatula Cake skewer Cooling rack Table knife
For frosting:				
2 oz	50 g	$\frac{1}{4}$ C	Butter or margarine	
4 oz	100 g	$\frac{1}{2}$ C	Cream cheese	
8 oz	225 g	2 C	Confectioner's sugar	
1 tsp	5 ml	1 tsp	Vanilla flavouring	

Variation For a carrot cake replace the same quantity of pumpkin with grated raw carrot and add 1 cup grated coconut or chopped nuts.

Method
1. Set the oven to 350°F and grease the baking pan.
2. Mix the eggs and sugar together. Beat in the vegetable oil.
3. Add the dry ingredients and mix well.
4. Add the pumpkin purée – made by boiling chopped pieces of pumpkin for 20 minutes, then strain and mash them.
5. Cook the mixture in the baking pan for 40 minutes, or until the mixture shrinks away from the side of the baking pan and a fine skewer comes out clean when inserted.
6. Prepare the frosting by creaming the butter and cream cheese together. Stir in the sifted confectioner's sugar and vanilla. Spread over the cooled cake.

Serve as a delicious dessert.

FOOD GROUP 4:
Food from animals

Foods from animals are important for their high quality protein and include meat, fish, eggs, milk and cheese. The recipes in this section are divided into these groups. Although they are usually more expensive than the other foods, only a small amount is required by the body. Protein is the most important nutrient for the growth and repair of all body cells.

Vitamins A and B and the minerals iron and calcium are also supplied by this group. Salt fish (salted cod) is a popular source of minerals including iodine. Almost perfect foods like milk and eggs have many uses in cookery and, as they are eaten by lacto-vegetarians, will be good sources of most of the nutrients. In tropical countries the sunshine provides all the vitamin D that is needed.

Animals used commonly for their meat in the Caribbean are hog (pork and bacon), beef and mutton (goat or sheep) and offal (the inside organs of an animal which can be eaten), e.g. liver and kidney. When buying meat avoid too much fat and, if the meat is frozen, allow it to thaw completely before cooking. This is especially important before seasoning poultry, such as whole or jointed chicken. Avoid overcooking these animal protein foods when using them as ingredients in the recipes.

MEAT

Meat Loaf

Ground or minced beef or hamburger has many uses for pies, pasta dishes, chillies, curries and an economical savoury meat loaf.

Quantity			Ingredients	Equipment
1	1	1	Onion – small size	Loaf pan – 2 lb/ 900 g size or shallow baking pan
2	2	2	Wholewheat bread slices	Sharp knife

Quantity			Ingredients	Equipment
1 lb	450 g	2 C	Ground beef/ hamburger	Cutting board
1	1	1	Egg – small size	Grater
Small amount			Salt and pepper	Mixing bowl
$\frac{1}{2}$ tsp	2 ml	$\frac{1}{2}$ tsp	Thyme – chopped	Wooden spoon
2 tblsp	30 ml	2 tblsp	Tomato paste	Teaspoon
1 tblsp	15 ml	1 tblsp	Hot pepper sauce	Tablespoon

Variation Incorporate a can of vegetable soup into the mixture.

Method
1. Set the oven to 325°F and grease the loaf pan.
2. Chop the onion and grate the slices of bread.
3. Mix the meat, onion and bread together in a bowl.
4. Mix in the egg, seasoning, tomato paste and sauce.
5. Pack into the loaf pan and bake for 45 minutes.

Serve turned out and garnished with salad when cool or place tomato slices and grated cheese on the top and bake for a further 5 minutes.

BBQ Ribs or Chicken

The unforgettable aroma and flavour of barbecued food is so appropriate for such a warm climate.

Quantity			Ingredients	Equipment
1 lb	450 g	2 C	Pork spare-ribs	Baking pan
4	4	4	or chicken pieces (whole cut up chicken)	Foil wrap Fruit squeezer
1	1	1	Lime	Sharp knife
Small amount			Salt and black pepper	Cutting board
1	1	1	Onion – small size	Tablespoon
$\frac{1}{4}$ pt	125 ml	$\frac{1}{2}$ C	Barbeque sauce – bottled or	
Home-made sauce:				
1	1	1	Onion	Cooking pot
1 oz	25 g	1 tblsp	Margarine or cooking oil	Wooden spoon
2 oz	50 g	2 tblsp	Brown sugar	

Quantity			Ingredients	Equipment
1 tblsp	15 ml	1 tblsp	Soya sauce or hot sauce	Measuring jug Can-opener
4 tblsp	60 ml	4 tblsp	Lime juice or vinegar	Teaspoon
$\frac{1}{4}$ pt	125 ml	$\frac{1}{2}$ C	Tomato ketchup	
1 tsp	5 ml	1 tsp	Cornstarch	
1	1	1	Pineapple with juice – 20 oz/550 g can size	

Variation Use a can of condensed tomato soup to make the sauce.

Method
1. Set the oven to 350°F and grease the baking pan.
2. Wash the meat and place in the baking pan. Cover with lime juice and seasoning.
3. Slice the onion and place over the meat.
4. Cover the baking pan with foil wrap and bake for 30 minutes.
5. Make the barbecue sauce by frying the chopped onion. Add the other ingredients and cook for 10 minutes. Mix in the cornstarch which has been mixed to a paste with some of the pineapple juice. Cook for a further 5 minutes.
6. Drain off any excess fat and pour the barbecue sauce generously over the meat.
7. Return to the oven uncovered for a further 10 minutes to brown.
   * The meat should be brown, moist and tender when cooked.

Serve as a main course meal with a staple food and vegetables.

Liver Casserole

A family 'one-pot' meal with many variations on flavour, with the addition of mushrooms, tomatoes, sweet peppers, hot peppers or curry powder.

Quantity			Ingredients	Equipment
1 lb	450 g	2 C –	Liver – lamb or pork	Casserole/stew dish
1 oz	25 g	$\frac{1}{4}$ C	Flour	Foil wrap
1 oz	25 g	1 tblsp	Margarine or cooking oil	Sharp knife

Food group 4: Food from animals 47

Quantity			Ingredients	Equipment
1	1	1	Onion	Cutting board
2	2	2	Carrots	Plastic bag
2	2	2	Celery sticks	Frying pan
1	1	1	Potato	Wooden spoon
1	1	1	Beef stew seasoning – packet (stock cube)	Measuring jug Tablespoon
$\frac{1}{2}$ pt	250 ml	1 C	Water	
Small amount			Salt and pepper	Teaspoon
$\frac{1}{2}$ tsp	2 ml	$\frac{1}{2}$ tsp	Mixed herbs or thyme	

Variation Use stewing steak, or chicken pieces if preferred, instead of liver.

Method
1. Set the oven to 350°F and wash the meat, removing any excess fat.
2. Cut the meat into small cubes and coat in the seasoned flour.
   * Use a plastic bag to do this.
3. Peel, wash and dice the vegetables.
4. Heat the fat and sauté the vegetables. Remove from the pan and add the meat to brown. Fry for 5 minutes and then place with the vegetables into the dish.
5. Dissolve the stock cube in the water, add to pan, with any remaining flour. Bring to the boil, stirring and add the seasoning.
6. Pour the gravy into the dish, cover and bake for $1-1\frac{1}{2}$ hours, until the meat is tender, or pressure cook on top of the stove.

Serve with dumplings and sprinkled with chopped parsley.

Hottie Patties

A favourite meat pie in parts of the Caribbean is a yellow coloured pastry envelope containing a seasoned meat mixture – not to be confused with beef burger patties

Quantity			Ingredients	Equipment
10 oz	275 g	$2\frac{1}{2}$ C	Flour	Baking pan
4 oz	100 g	$\frac{1}{2}$ C	Margarine or butter	Mixing bowl

Quantity			Ingredients	Equipment
1 tsp	5 ml	1 tsp	Baking powder	Sieve
Small amount			Yellow food colouring	Measuring jug
To mix			Water	Table knife
Filling:				
8 oz	225 g	1 C	Ground beef/ hamburger	
$\frac{1}{2}$	$\frac{1}{2}$	$\frac{1}{2}$	Onion	
1-2	1-2	1-2	Hot peppers – small size	Cooking pot Wooden spoon
For frying			Cooking oil	
1	1	1	Beef stew seasoning – packet (stock cube)	Tablespoon Teaspoon Rolling pin Flour dredger Saucer or plate – small size
Small amount			Salt and pepper	
$\frac{1}{2}$ tsp	2 ml	$\frac{1}{2}$ tsp	Thyme – chopped	Fork

Variation Use a can of corned beef for the meat and for hotter patties use curry powder or hot sauce to season.

Method
1. Make the pastry by rubbing the margarine into the flour, using the fingertips.
2. Add a few drops of the food colouring to the water and use to bind the dough together. Chill in the refrigerator.
3. Chop the onion and hot peppers and sauté in the cooking oil. Brown the ground beef and add the seasoning with a little water. Simmer for 5–10 minutes.
4. Set the oven to 400°F and grease the baking pan.
5. Roll out the pastry and cut approximately 6 circles, using a saucer as a guide.
6. Place the filling on to half of the pastry circle, fold over, dampen the edges with water and seal with a fork. Bake for 30 minutes until brown.

Serve wrapped in foil wrap to keep warm for lunch.

Pineapple Pork Chops

Meat and fruit are complementary, reducing any richness of flavour and providing a perfect colour coordination.

Food group 4: Food from animals 49

Quantity			Ingredients	Equipment
4	4	4	Pork chops	Baking dish
4	4	4	Pineapple slices – small can size	Foil wrap
1 tblsp	15 ml	1 tblsp	Lime juice	Can-opener
2 tblsp	30 ml	2 tblsp	Brown sugar	Tablespoon

Variation Try pork chops with apple rings or chicken pieces with orange slices.

Method
1. Set the oven to 400°F and grease the baking dish.
2. Wash and season the pork chops. Place in the dish, laying pineapple slices on top.
3. Add about 4 tablespoons of the canned pineapple juice and the lime juice.
4. Sprinkle the sugar on top, cover with foil wrap and bake for 30 minutes. Uncover to brown for 10 minutes before serving.

Serve with a staple food and green vegetables or salad.

FISH

Baked Grouper Fish

Make use of a wide range of available fish to produce an appetising meal, seasoned with locally produced hot sauce – of which the temperature varies with each individual island.

Quantity			Ingredients	Equipment
1 lb	450 g	2 C	White fish, e.g. grouper, red snapper, king fish	Baking pan
1	1	1	Lime	Foil wrap Sharp kinfe
Small amount			Pepper and thyme	Cutting board
1	1	1	Onion – small size	Fruit squeezer
1	1	1	Sweet pepper – small size	Tablespoon Can-opener
1	1	1	Celery stick	Grater
8 oz	225 g	1 C	Tomatoes – fresh or canned	
1 tblsp	15 ml	1 tblsp	Hot sauce	

Quantity			Ingredients	Equipment
1-2	1-2	1-2	or hot peppers	
2	2	2	Wholewheat bread slices	

Variation Add chopped mushrooms or canned sweet-corn to this savoury sauce.

Method
1. Set the oven to 350°F and grease the baking pan.
2. Wash and clean the fish. Remove the skin and debone if necessary.
3. Season the fish and place in the baking pan with the lime juice. Allow to marinate.
4. Finely chop the vegetables and add to the fish with the hot sauce. Add a small amount of water if using fresh tomatoes.
5. Grate the bread and sprinkle the breadcrumbs on top.
6. Cover with foil wrap and bake for 30 minutes. Remove the cover and continue baking for a further 10 minutes.

Serve sprinkled with grated Parmesan cheese and freshly cooked green vegetables.

Island Fish Pie

A quick supper dish combining a staple food, fish, which can be either fresh or canned, and the use of coconut milk for a real island taste.

Quantity			Ingredients	Equipment
2	2	2	Potatoes – large size	Baking dish
8 oz	225 g	1 C	Fish – fresh or canned, e.g. tuna, jack	Cooking pot Potato peeler
$\frac{1}{2}$	$\frac{1}{2}$	$\frac{1}{2}$	Onion	Sharp knife
2 oz	50 g	$\frac{1}{4}$ C	Cheese – grated	Cutting board Grater
1 oz	25 g	1 tblsp	Butter or margarine	Potato masher Can-opener
3 tblsp	45 ml	3 tblsp	Coconut milk or cream	Fork
Small amount			Salt and pepper	Tablespoon
1	1	1	Tomato	Sieve

Variation Use noodles as an alternative staple food and add a small amount of tomato paste to the mixture.

Method
1. Set the oven to 350°F and grease the baking pan.
2. Peel the potatoes and cook for 20 minutes in a small amount of water.
   * If using fresh fish, place on a plate to steam above the boiling potatoes.
3. Finely chop the onion and grate the cheese.
4. Mash the potatoes and flake the fish.
5. Mix all the ingredients together, adding the milk to obtain a soft consistency. Half of the grated cheese can be saved for sprinkling on top.
6. Garnish the top with the sliced tomato and bake for 30 minutes until golden brown.

* Make the coconut milk by adding water/milk to grated coconut meat. Stand for an hour and squeeze out the liquid through a sieve.

Serve warm with salad and garnished with cooked shrimps.

Fish Fingers

Salt fish fritters make a breakfast dish or are more commonly served with peas and rice for a main meal.

Quantity			Ingredients	Equipment
12 oz	350 g	$\frac{3}{4}$ C	Fish – white flesh, e.g. grouper, snapper	Serving dish
2 oz	50 g	$\frac{1}{2}$ C	Flour	Sharp knife
Small amount			Salt and pepper	Cutting board
1	1	1	Egg	Plates – 3
$2\frac{1}{2}$ oz	65 g	$\frac{1}{2}$ C	Cornmeal or cracker meal	Fork
For frying			Cooking oil	Frying pan
Small amount			Parmesan cheese	Draining spoon
1	1	1	Lime	Kitchen paper
To taste			Hot sauce	Teaspoon

Variation Use a thick batter to coat the fish (as for frittered fruit) or steam first and make into cakes with mashed potato and bind together with an egg before frying.

Method
1. Wash, fillet and skin the fish if necessary. Cut into slices $1\frac{1}{2}$ ins/$3\frac{1}{2}$ cm wide.
2. Dip in seasoned flour and then into beaten egg. The fish can also be seasoned with a little lime juice and hot sauce to taste.
3. Coat the fish with cornmeal and fry in deep hot cooking oil for 5 minutes until golden brown.
   * Avoid getting the fat too hot and test with a small cube of bread before frying.
4. Remove the fish and drain on kitchen paper. Sprinkle with Parmesan cheese and place on a serving dish.

Serve with lime wedges, hot sauce and some tartare sauce.

Shellfish Salad

An ideal way of using fresh seafood or cooked meat, with tomatoes, sweet pepper, sour lime and mayonnaise.

Quantity			Ingredients	Equipment
1 lb	450 g	2 C	Shellfish, e.g. conch, crab or canned fish, e.g. tuna	Serving bowl with cover / Can-opener / Fork
$\frac{1}{2}$	$\frac{1}{2}$	$\frac{1}{2}$	Onion – small size	Sharp knife
$\frac{1}{2}$	$\frac{1}{2}$	$\frac{1}{2}$	Sweet pepper	Cutting board
1	1	1	Celery stick	Fruit squeezer
2	2	2	Tomatoes	Tablespoon
$\frac{1}{2}$	$\frac{1}{2}$	$\frac{1}{2}$	Cucumber	Teaspoon
1	1	1	Lime	
1–2	1–2	1–2	Hot peppers – small size	

Variation Use cooked chicken instead of fish, omit the lime and add 2 tablespoons of mayonnaise and 1 teaspoon of mustard.

Method
1. Dice or flake the cleaned fish.
2. Peel and finely chop the onion with the other vegetables.
3. Squeeze the juice from the lime and combine all the ingredients together. Allow to marinate, covered in the refrigerator, before serving.

Serve garnished on a bed of shredded crisp lettuce.

Fish Chowder

A juicy thick soup, traditionally served with sherry or layered with lightly crushed biscuits and sliced cooked potatoes.

Quantity			Ingredients	Equipment
1 lb	250 g	2 C	Fish, e.g. conch, crawfish or white fish	Flask or container with cover
1	1	1	Onion	Sharp knife Cutting board
2	2	2	Carrots	Cooking pot – large size
1	1	1	Potato	
2 oz	50 g	$\frac{1}{4}$ C	Salt pork or cooking oil	Wooden spoon Can-opener
8 oz	225 g	1 C	Tomatoes – fresh or canned	Tablespoon
1 tblsp	15 ml	1 tblsp	Tomato paste	Teaspoon
Small amount			Salt and pepper	
$\frac{1}{2}$ tsp	2 ml	$\frac{1}{2}$ tsp	Thyme – chopped	
1	1	1	Bay leaf	
To taste			Hot pepper sauce	

Variation Use turtle meat to make a chowder.

Method
1. Wash the fish and chop (or grind the conch).
2. Peel and dice the onion, carrots and potato.
3. Chop the salt pork and heat in a cooking pot. Sauté the vegetables for 5 minutes.
4. Add the chopped tomatoes, tomato paste, seasoning and the prepared fish. Add a small amount of water if using fresh tomatoes.
5. Bring to the boil and reduce the heat to a simmer for 40 minutes. Remove any remaining salt pork before serving.

Serve with a squeeze of lime juice and hot buttered Johnny cake.

EGGS

Cup Cakes

The basic 'creaming' method cake mix, which can be adapted to batch-bake small buns, coconut cake, cookies, fruit upside-down cake and iced birthday layer cake.

Quantity			Ingredients	Equipment
4 oz	100 g	½ C	Margarine or butter	Cup cake pan or 2 round cake pans
4 oz	100 g	½ C	Sugar	Paper cup cake cases
2	2	2	Eggs	Mixing bowl
4 oz	100 g	1 C	Flour – all-purpose	Wooden spoon
1 tsp	5 ml	1 tsp	Baking powder	Sieve
½ tsp	2 ml	½ tsp	Vanilla flavouring	Teaspoon
Small amount			Yellow food colouring – optional	Tablespoon

For frosting:

4 oz	100 g	¼ C	Butter	Spatula
8 oz	225 g	2 C	Confectioner's sugar	Cooling rack
Small amount			Milk	Table knife

Variation Add a different flavouring to the basic mixture, e.g. raisins, chopped nuts, grated coconut, glacé cherries or chocolate chips.

Method
1. Set the oven to 375°F and grease the baking pan or line the cup cake pan with the cup cake cases.
2. Cream the margarine and sugar together in a mixing bowl, using a wooden spoon, until the mixture is pale, soft and fluffy.
3. Gradually add the beaten eggs, with a little of the sifted flour.
4. Fold in the remaining flour and baking powder lightly with a tablespoon.
   * The consistency should be 'soft dropping' – add a little milk if necessary.
5. Fill the cup cake cases two-thirds full or the round pans half full. Level the surface and bake for 20–25 minutes until golden brown and firm to touch.
6. Turn cakes on to a cooling rack. Prepare the icing by creaming together the butter and confectioner's sugar, or for a lighter icing make up the confectioner's sugar with water.

Serve a sandwich cake by spreading jam or jelly in between two cake layers and sprinkle the top with sugar.

Jelly Roll

A very light sponge, skilfully made from whisked eggs and filled with a home-made preserve or lemon curd.

Quantity			Ingredients	Equipment
2	2	2	Eggs	Baking pan – 7 × 12 ins/18 × 30 cm size
2 oz	50 g	$\frac{1}{4}$ C	Sugar	Foil wrap or waxed paper
2 oz	50 g	$\frac{1}{2}$ C	Flour	Mixing bowl
1 tblsp	15 ml	1 tblsp	Water – hot	Whisk
Small amount			Confectioner's sugar	Sieve
				Tablespoon
				Spatula
3 tblsp	45 ml	3 tblsp	Fruit preserve, e.g. guava jelly, orange marmalade, lemon curd	Sharp knife Clean drying cloth Cooling rack

Variation Make a larger size roll with 3 eggs to 3 oz/75 g sugar and flour (baking pan – 9 × 12 ins/22$\frac{1}{2}$ × 30 cm). Use fresh fruit purée and whipped cream to fill.

Method
1. Set the oven to 400°F and grease and line the base of the baking pan.
2. Whisk the eggs with the sugar until thick and pale in colour.
   * The mixture should hold the trail of the beater.
3. Gently fold in the sieved flour and water using a tablespoon.
4. Spread into the baking pan and bake for 8–10 minutes, avoiding over-cooking.
5. Have ready a damp drying cloth, covered with a sheet of waxed paper and sprinkled finely with sugar.
6. When the cake is ready turn on to the sugared paper. Peel off the paper lining, trim the edges and spread quickly with warmed jelly.
7. Make a small cut across the cake at one end and roll up, using the paper to help. Cool on a cooling rack.

Serve cut into slices. This makes an excellent trifle dessert base.

Potato Salad

An excellent way of using eggs as an accompaniment for many meat and fish dishes. This salad is often served at picnics and barbecues.

Quantity			Ingredients	Equipment
3	3	3	Eggs	Plastic container with cover
3	3	3	Potatoes – large size	Cooking pots – 2
$\frac{1}{2}$	$\frac{1}{2}$	$\frac{1}{2}$	Onion	Sharp knife
1	1	1	Celery stick	Cutting board
2 tblsp	30 ml	2 tblsp	Mayonnaise	Mixing bowl
1 tsp	5 ml	1 tsp	Mustard	Teaspoon
1 tblsp	15 ml	1 tblsp	Lime juice	Tablespoon
Small amount			Salt and pepper	

Variation Use pasta shapes instead of potatoes and make your own mayonnaise.

Method
1. Boil the eggs for 10 minutes and cool in cold water when ready.
2. Peel and quarter the potatoes. Boil for 15 minutes until just tender.
3. Finely chop the onion and celery and combine with the remaining ingredients in a large bowl.
4. Drain, cool and dice the potatoes.
5. Peel and chop two of the eggs and slice the other.
6. Add the potato and chopped egg to the mayonnaise mixture and toss lightly. Spoon into a serving bowl.

Serve garnished with sliced egg and some chopped sweet green pepper.

Lemon Cheese

Using available citrus fruit, the secret behind this recipe is not to let the mixture get too hot, otherwise the egg whites will cook first.

Quantity			Ingredients	Equipment
3	3	3	Lemons/limes or use oranges	Jars and covers – small size

Quantity			Ingredients	Equipment
				Grater
				Fruit squeezer
4 oz	100 g	½ C	Butter or margarine	Double boiler or cooking pot and small bowl
8 oz	225 g	1 C	Sugar	Measuring jug
3	3	3	Eggs – whole or yolks only	Fork Wooden spoon

Variation Use this mixture to fill a pastry pie case and top with whisked egg whites to make a *Lemon Meringue Pie*.

Method
1. Finely grate a small amount of rind from the washed lemons and squeeze the juice.
2. Place the butter, sugar, lemon rind and juice in a double boiler or small bowl placed over a cooking pot of boiling water.
3. Beat the eggs with a fork and add to the mixture. Stir with a wooden spoon, keeping the water in the cooking pot just below boiling point.
4. Cook for about 5 minutes until thick. Pour into clean, warm, small jars. Seal, label and store in the refrigerator.

*Seal with a screw top or other lid to obtain a vacuum within the jar or bottle. The lid will then remain firm when tested.

Serve spread on freshly baked scones or bread, or as an alternative filling for a layer cake.

Supper Sandwiches

'*Croque Monsieur*' in the French-speaking Caribbean, this carefully fried nutritious snack can be eaten straight away for lunch while still warm.

Quantity			Ingredients	Equipment
1 oz	25 g	1 tblsp	Butter or margarine	Serving plate Foil wrap
4	4	4	Wholewheat bread slices	Sharp knife
2	2	2	Ham – cooked slices	Cutting board

Quantity			Ingredients	Equipment
2	2	2	Cheese slices	Table knife
1	1	1	Egg	Measuring jug
2 fl.oz	60 ml	$\frac{1}{4}$ C	Milk	Fork
To season			Paprika pepper	Frying pan
For frying			Cooking oil	Draining spoon
				Kitchen paper

Variation Another slice of bread can be used to make a 'double-decker' effect of bread/ham/bread/cheese/bread. Make up your own filling, e.g. corned beef and tomato or cooked bacon and mushroom.

Method
1. Butter one side of the bread and lay a slice of ham and cheese in between the buttered slices of bread, to make 2 sandwiches. Press together and cut into triangles.
2. Beat the egg and milk together with a fork, season and pour on to a plate. Dip the sandwiches in to absorb the mixture.
3. Heat the cooking oil in a frying pan and quickly fry the sandwiches, making sure that the pan does not get too hot. Brown on each side.

Serve garnished with green grapes or with a watercress or lettuce salad.

MILK

Caribbean Quiche

An appetising savoury pie which makes use of several dairy products. The rest of the filling can be adapted according to what is in store.

Quantity			Ingredients	Equipment
6 oz	150 g	$1\frac{1}{2}$ C	Flour	Pie dish – 9 ins/ 23 cm
3 oz	75 g	$\frac{1}{3}$ C	Margarine	Mixing bowl
To mix			Water	Sieve
$\frac{1}{2}$	$\frac{1}{2}$	$\frac{1}{2}$	Onion	Measuring jug
$\frac{1}{2}$	$\frac{1}{2}$	$\frac{1}{2}$	Sweet pepper	Table knife
For frying			Cooking oil	Rolling pin
$\frac{1}{4}$ pt	$\frac{1}{4}$ pt	$\frac{1}{4}$ pt	Milk	Flour dredger

Quantity			Ingredients	Equipment
2	2	2	Eggs	Sharp knife
1	1	1	Tuna fish – small can size	Cutting board
2 oz	50 g	¼ C	Cheese – grated	Frying pan – small size
1	1	1	Tomato	Wooden spoon
To season			Black pepper	Fork
				Teaspoon
½ tsp	2 ml	½ tsp	Mixed herbs or parsley	Can-opener
				Grater

Variation Make wholewheat pastry, using half quantity wholewheat flour and add any of the following to the filling: chopped mushrooms, cooked ham or bacon, chicken or sweetcorn.

Method
1. Set the oven to 375°F and grease the pie dish.
2. Make short-crust pastry by rubbing the margarine into the flour and slowly adding the water. Roll out to line the pie dish.
3. For the filling chop the onion and sweet pepper and sauté in the cooking oil until soft, but not brown. Place into the pastry case.
4. Beat the milk and eggs together, adding the seasoning.
5. Place flaked tuna fish into the pastry case and pour over the egg mixture.
6. Grate the cheese and sprinkle over the surface. Garnish with sliced tomato and bake for 30–40 minutes, until firm to touch.

Serve either hot or cold with a crisp green salad.

Tossed Pancakes

Pancakes make an economical and nutritious main course dish filled with a seafood sauce or savoury minced beef filling, and may be served as warmed dessert crêpes filled with fruit.

Quantity			Ingredients	Equipment
4 oz	100 g	1 C	Flour	Serving plate
				Foil wrap
Small amount			Salt	

Quantity			Ingredients	Equipment
1	1	1	Egg	Mixing bowl
½ pt	250 ml	1 C	Milk	Sieve
For frying			Cooking oil	Wooden spoon or whisk
				Measuring jug
1	1	1	Lemon or lime	Tablespoon
2 oz	50 g	¼ C	Sugar	Frying pan – small size
				Cup – small size
				Table knife
				Kitchen paper

Variation Add ½ teaspoon of baking powder to the batter and use to make small dropped scones, on the hot surface of a frying pan.

Method
1. Sieve the flour and salt into a mixing bowl.
2. Add the egg and half the milk. Beat until smooth.
3. Mix in the remaining milk and pour into a jug.
   * To prevent the pancakes from sticking, 1 tablespoon of the cooking oil can be added to the batter.
4. Heat the oil in a frying pan. When the fat is hot, pour excess into a cup at the side.
5. Pour in enough batter to cover bottom of the pan. Cook for 2–3 minutes until the underside of the pancake is golden brown. Toss or turn it over with a knife to cook the other side.
6. Put on to a plate and repeat the process using remaining batter. If the pancakes are to be served straight away, keep warm in foil wrap in the oven.

Serve individually rolled up to eat, sprinkled with sugar and with a squeeze of lime juice. Alternatively fill with canned fruit pie filling or syrup.

Chocolate Pudding Pie

An irresistible smooth rich dessert in a biscuit crumb crust which can be served with a meringue or whipped cream topping.

Quantity			Ingredients	Equipment
6 oz	150 g	1½ C	Biscuits – graham cracker or digestive	Pie dish – 9 ins/ 23 cm size
3 oz	75 g	⅓ C	Margarine or butter	Plastic bag

Food group 4: Food from animals

Quantity			Ingredients	Equipment
1 oz	25 g	1 tblsp	Cornstarch	Rolling pin
2 tblsp	30 ml	2 tblsp	Cocoa	Cooking pot
$\frac{3}{4}$ pt	375 ml	$1\frac{1}{2}$ C	Milk	Wooden spoon
1 oz	25 g	2 tblsp	Margarine	Measuring jug
2 oz	50 g	$\frac{1}{4}$ C	Sugar	Tablespoon
To decorate			Chocolate	Grater
			Nuts – chopped	
			Whipping cream	

Variation Make a short-crust pastry pie base and top with meringue.

Method
1. Melt the margarine in a cooking pot, using a little of it to grease the pie dish.
2. Crush the biscuits by placing them in a plastic bag and using a rolling pin.
3. Add the crumbs to the margarine in the cooking pot and mix well. Use to line the sides and base of the pie dish.
4. Blend the cornstarch and cocoa with 2 tablespoons of the milk. Heat the rest of the milk.
5. When milk boils pour it on to the cornstarch, stirring all the time. Return mixture to the cooking pot, bring to the boil and stir with a wooden spoon.
6. Add the margarine and sugar and beat until smooth. Pour the filling into the pie case.

Serve cold sprinkled with grated chocolate and chopped nuts around the edges.

Pawpaw Nectar

Frothy milk-shakes or this high vitamin C drink made from the sweet and juicy ripe pulp of the pawpaw, also known as papaya, are delicious.

Quantity			Ingredients	Equipment
1	1	1	Pawpaw – ripe – large size	Plastic bottle or drink container
$\frac{1}{2}$ pt	250 ml	1 C	Evaporated milk or single cream	Sharp knife
2	2	2	Limes – large size	Cutting board
				Teaspoon

Quantity			Ingredients	Equipment
				Potato masher
4 oz	100 g	½ C	Sugar	Fruit squeezer
				Grater
½ tsp	2 ml	½ tsp	Vanilla flavouring	Can-opener
				Measuring jug
8 tblsp	120 ml	1 C	Ice – finely crushed	Electric blender or hand whisk

Variation Watermelon, mango, guava and sugar or custard apple also make a good nectar.
* If using hard fruit, cook first until tender.

Method
1. Peel the pawpaw, cut in half lengthwise and scoop out the seeds. Chop into pieces or mash.
2. Squeeze the juice from the limes and finely grate 1 teaspoon of the rind.
3. Place all ingredients into an electric blender for 30 seconds, or whisk the mixture by hand, until smooth and thick.

Serve as an appetiser in glasses over ice and decorated with thin slices of lime.

Natural Yogurt

Yogurt makes a healthy ingredient for salad dressings, as a side-dish for hot and spicy meat dishes and served flavoured according to individual taste with fresh fruit or chopped nuts.

Quantity			Ingredients	Equipment
1 pt	500 ml	2 C	Evaporated milk (made up with water) or sterilised milk	Vacuum flask
1 tblsp	15 ml	1 tblsp	Natural yogurt	Cooking pot – large size
				Can-opener
1 tblsp	15 ml	1 tblsp	Dried milk powder	Measuring jug
				Thermometer
				Tablespoon
				Whisk
				Measuring jug

Variation Mix completed yogurt with chopped fresh fruit, melted chocolate or lemon curd preserve.

Method
1. Sterilise with boiling water the vacuum flask and all equipment to be used.
2. Measure the milk into the cooking pot and heat until just warm to touch with a finger.
   * The temperature is 112°F/44°C with a thermometer.
3. Whisk in the natural yogurt and dried milk powder.
4. Pour into the vacuum flask and leave to set for at least 8 hours. Transfer to covered containers and store in the refrigerator.

Serve with puddings and desserts instead of cream. Add a small amount of honey for sweet yogurt.
   * If using fresh pasteurised milk or untreated milk, boil the milk first and leave to cool down until warm.

CHEESE

Cheese and Ham Roll-Ups

A tasty combination of bananas wrapped in ham and covered with a cheese sauce, shows skill and the successful use of dairy products.

Quantity			Ingredients	Equipment
2 tsp	10 ml	2 tsp	Mustard	Baking dish
4	4	4	Cooked ham slices	Teaspoon
4	4	4	Bananas – small size	Table knife
$\frac{1}{2}$	$\frac{1}{2}$	$\frac{1}{2}$	Lime	Cutting board
1 oz	25 g	1 tblsp	Margarine or butter	Cooking pot
1 oz	25 g	2 tblsp	Flour	Tablespoon
$\frac{1}{2}$ pt	250 ml	1 C	Milk	Measuring jug
Small amount			Salt and pepper	Wooden spoon
4 oz	100 g	$\frac{1}{2}$ C	Cheese – grated	Grater

Variation Roll up coleslaw salad or spinach inside the ham, instead of bananas.

Method
1. Spread the prepared mustard on to the ham slices and place one banana on each. Sprinkle with lime juice and roll up.

2. Prepare the sauce: Heat the margarine and add the flour. Cook for a few minutes on a low heat. Slowly add the milk off the heat. Season with salt and pepper and continue stirring with a wooden spoon on low heat.
3. Grate the cheese and add two-thirds of this off the heat to the thickened sauce.
4. Pour the sauce over the ham rolls, which have been arranged in a greased baking dish, and sprinkle with the remaining cheese. Brown under a hot broiler/grill or in the oven.

Serve hot with sliced tomatoes and wholewheat bread.

Macaroni Cheese

Cut into wedges, this pasta dish is served like a cake with a main course dish or packed up for lunch.

Quantity			Ingredients	Equipment
8 oz	225 g	2 C	Macaroni	Baking pan – 9 × 12 ins/22½ × 30 cm size
1	1	1	Onion – small size	Cooking pot
½	½	½	Sweet pepper – small size	Sharp knife
1	1	1	Celery stick	Cutting board
8 oz	225 g	1 C	Cheese – grated	Grater
2	2	2	Eggs	Measuring jug
1	1	1	Evaporated milk – 14 oz/400 g can size	Fork Can-opener Strainer
2 oz	50 g	¼ C	Margarine or butter	Tablespoon
Small amount			Salt and pepper	

Variation Mix the boiled macaroni into a cheese sauce made from a 'roux'.

Method
1. Set the oven to 350°F and grease the baking pan.
2. Half fill a cooking pot with water and add the macaroni. Bring to the boil and simmer for 10 minutes.
3. Chop vegetables finely and grate the cheese.

4. Beat the eggs in a measuring jug and add the evaporated milk and seasoning.
5. Drain the macaroni well and add all the other ingredients. Bake for 30-40 minutes, until golden brown on top.

Serve with a sliced tomato or coleslaw salad.

Cheese Pizza

Enjoyable to eat and to make into interesting seasonal shapes, for example a Christmas tree, pumpkin, oblong or individual rounds to fit the plate.

Quantity			Ingredients	Equipment
8 oz	225 g	2 C	Flour – whole-wheat or all-purpose	Pizza pan
1 tsp	5 ml	1 tsp	Baking powder	Mixing bowl
2 oz	50 g	$\frac{1}{4}$ C	Margarine or butter	Sieve
$\frac{1}{4}$ pt	125 ml	$\frac{1}{2}$ C	Milk	Teaspoon
4 oz	100 g	$\frac{1}{2}$ C	Cheese – grated	Table knife
				Measuring jug
$\frac{1}{2}$	$\frac{1}{2}$	$\frac{1}{2}$	Onion	Rolling pin
$\frac{1}{2}$	$\frac{1}{2}$	$\frac{1}{2}$	Sweet pepper	Flour dredger
1	1	1	Tomato paste – small can	Grater
Small amount			Pepper and mixed herbs	Sharp knife

Choose an additional topping:

2 oz	50 g	$\frac{1}{4}$ C	Mushrooms	Cutting board
4 oz	100 g	$\frac{1}{2}$ C	Bacon	Can-opener
4 oz	100 g	$\frac{1}{2}$ C	Sausage, e.g. pepperoni	
1	1	1	Sardines – can	
1	1	1	Olives – small bottle	

Variation Use a yeasted bread dough or purchased muffins for the base.

Method
1. Set the oven to 400°F and grease the baking pan.

2. Sieve the flour and baking powder into a mixing bowl and rub in the margarine, using the fingertips.
3. Slowly add the milk and mix to a dough.
4. Roll out or flatten with the hands into a large circle (about 10 ins in diameter).
5. Grate cheese and chop vegetables.
6. Spread tomato paste, cheese, seasoning and additional ingredients over the base.
7. Bake for about 30 minutes, until the base is cooked through.

Serve warm or cold, cut into slices with salad.

Simple Cheese-Cake

Cheese-cakes are an extra special dessert, made either light and creamy style or with a heavier baked texture – they deserve second helpings all around!

Quantity			Ingredients	Equipment
6 oz	150 g	1½ C	Biscuits – graham cracker or digestive	Pie dish – 8 ins/ 20 cm size
3 oz	75 g	⅓ C	Margarine or butter	Plastic bag
1	1	1	Lemon or lime	Rolling pin
¼ pt	125 ml	½ C	Whipping cream	Cooking pot – small size
8 oz	225 g	1 C	Cottage cheese or cream cheese	Wooden spoon
2 oz	50 g	¼ C	Sugar	Grater
8 oz	225 g	1 C	Fruit, e.g. Cherries, guavas, oranges or canned fruit pie filling	Fruit squeezer Mixing bowl Whisk Sieve Tablespoon

Variation Use lemon slices and piped cream to decorate, instead of sliced fresh fruit.

Method
1. Make the biscuit crumbs by placing the biscuits in a plastic bag and crushing them with a rolling pin.
2. Melt the margarine and stir in the crumbs. Line the base and sides of the pie dish. Chill in the refrigerator.

3. Finely grate the rind from the washed lemon and squeeze the juice.
4. Whisk the cream until thick and add the sieved cottage cheese. Fold in the sugar and lemon rind and juice.
5. Spread the mixture over the crumb base and return to the refrigerator.

Serve decorated with fresh Caribbean fruit of your own choice.

Cheese Shortbread

A traditional shortbread recipe makes a useful appetiser, often called cheese straws, which are made by the rubbing-in method.

Quantity			Ingredients	Equipment
6 oz	150 g	$1\frac{1}{2}$ C	Flour – wholewheat or all-purpose	Baking pan – flat size
4 oz	100 g	$\frac{1}{2}$ C	Butter or margarine	Mixing bowl
2 oz	50 g	$\frac{1}{4}$ C	Cheese – grated	Grater
$\frac{1}{4}$ tsp	1 ml	$\frac{1}{4}$ tsp	Baking powder	Teaspoon
				Table knife
Small amount			Paprika or cayenne pepper	Rolling pin
				Flour dredger
				Cooling rack

Variation Add a tablespoon of tomato ketchup for flavour and to help bind the mixture together.

Method
1. Set the oven to 350°F and grease the baking pan.
2. Rub the butter into the flour using the fingertips.
3. Add the grated cheese and baking powder and mix together with the hands.
4. Roll out the dough on a lightly floured work surface. Cut into 'fingers' ($\frac{1}{4}$ ins/$\frac{1}{2}$ cm thick) and place on to the baking pan.
5. Sprinkle with paprika and bake for 20 minutes, until beginning to brown. Cool on a cooling rack and store in an airtight container.

Serve with your favourite dip, e.g. soured cream flavoured with onion or avocado.

FOOD GROUP 5:
Fruits

Fruits differ from vegetables in that they contain seeds and are commonly used in salads or the dessert part of the menu. There is a wide selection of local fruits available, e.g. guava, papaya, pineapple, mango, soursop, grapefruit and banana.

Fruits help to provide the essential vitamins and minerals required to keep our bodies free from disease. Citrus fruits, e.g. orange and lime, and the acid-fruits, e.g. guava, pineapple and West Indian cherries contain vitamin C. Encourage the eating of these fruits and green vegetables in meals, to assist in the absorption of iron.

Bright yellow coloured fruits, e.g. papaya and mango contain vitamin A in the form of carotene. In addition to a small quantity of minerals, fruits also contain carbohydrates (sugar or starch) and water. The fibre content in fruit is softened by cooking and so it is preferable to eat them raw, for example in a fresh fruit salad recipe.

Preserve fruits, so that they can be used in recipes out of season, by bottling soft fruit and making jams, jellies or marmalades from citrus fruit (see recipes for *Lemon Cheese* and *Mixed Fruit Marmalade*), and drying fruit such as apples, grapes and plums (see recipe for *Sweet Mincemeat*). Make use of spices, which are the dried seeds and berries from tropical plants, e.g. ginger and nutmeg, sold whole or ground, to flavour these dishes.

Fresh Fruit Cocktail

This delicious syrup base is poured over a colourful combination of fresh Caribbean fruit, which can be displayed in a scooped out pineapple or melon shell for a special occasion.

Quantity			Ingredients	Equipment
4 oz	100 g	½ C	Sugar	Plastic container with cover
1 pt	500 ml	2 C	Water	Cooking pot – small size
1	1	1	Lime or lemon	Vegetable peeler

Quantity	Ingredients	Equipment
		Sharp knife
		Cutting board
		Fruit squeezer
		Strainer or sieve

Choose at least one fruit from each group:

Citrus	Hard	Soft	Stone	Exotic
Orange	Apple	Banana	Cherry	Pineapple
Grapefruit	Pear	Grape	Mango	Guava
Tangerine		Strawberry	Plum	Papaya
Kumquat		Sugar apple	Peach	Passion
		Melon	Apricot	fruit
				Carambola

Variation   Use chunks of fruit also for skewered kebabs or added to a cake mixture to produce a fruit cocktail cake.

Method
1. Make the syrup by dissolving the sugar in the water in a cooking pot, on a low heat.
2. Peel the 'zest' from the washed citrus fruits and add to the syrup.
3. Wash and cut up all the fruits. Remove seeds, cores and skins as necessary.
4. Squeeze the lime or lemon and use the juice to cover the fruits, to prevent browning.
5. Add the strained cooled syrup to the mixed fruits.

Serve with natural yogurt or ice-cream.

Lattice Fruit Pie

Variations of fruit fillings for baked pies are endless, ranging from classic apple to crushed pineapple, bananas, or sliced unripe papaya.

Quantity	Ingredients	Equipment
1 lb 450 g 2 C	Fruit, e.g. Pineapple, bananas, cherries, papaya	Pie dish – 9 ins/ 23 cm size

Quantity			Ingredients – fresh or canned	Equipment
2–3 oz	50–75 g	$\frac{1}{4}-\frac{1}{3}$ C	Sugar – brown	Cooking pot
$\frac{1}{2}$ tsp	2 ml	$\frac{1}{2}$ tsp	Cinnamon	Sharp knife
$\frac{1}{2}$	$\frac{1}{2}$	$\frac{1}{2}$	Lime or lemon	Cutting board
8 oz	225 g	2 C	Flour	Teaspoon
8 oz	225 g	2 C	Flour	Can-opener
4 oz	100 g	$\frac{1}{2}$ C	Margarine or butter	Mixing bowl Sieve
To mix			Water	Measuring jug Table knife
To glaze			Milk	Rolling pin Flour dredger Tablespoon Pastry brush

Variation Using a little more pastry roll out to cover the whole surface instead of the lattice of pastry strips. Add some dried raisins to the fruit mixture.

Method
1. Wash and slice fresh fruit and place in a cooking pot with sugar, cinnamon, lime juice and a small quantity of water. Bring to the boil and simmer for 10 minutes.
   * There is no need to cook bananas.
2. Set the oven to 400°F and grease the pie dish.
3. Prepare short-crust pastry by placing sifted flour into a mixing bowl and rubbing in the margarine using fingertips.
4. Slowly add the water and mix to a stiff dough consistency.
5. Divide the pastry into two equal pieces. Roll out one to line the pie dish.
6. Place fruit mixture into the pastry case. Dampen edges of pastry with water.
7. Roll out remaining pastry, cut out strips and make a lattice over the fruit filling. Seal outer edges together.
8. Brush with a little milk and sprinkle with some extra sugar. Bake for 35 minutes until the crust is crisp and golden brown.

Serve warm with ice-cream.

Sugared Fruit Fritters

Frittered fruit makes a good accompaniment to ham and chicken dishes and the basic batter can be used for coating fish and other fried foods.

Food group 5: Fruits

Quantity			Ingredients	Equipment
4 oz	100 g	1 C	Flour	Serving plate
1	1	1	Egg	Can-opener
$\frac{1}{4}$ pt	125 ml	$\frac{1}{2}$ C	Milk	Mixing bowl
For frying			Cooking oil	Whisk
1 oz	25 g	1 tblsp	Sugar	Tablespoon
$\frac{1}{2}$ tsp	2 ml	$\frac{1}{2}$ tsp	Cinnamon or ginger	Frying pan
			Fruit,	Draining spoon
			e.g. Pineapple, bananas, mandarin oranges	Kitchen paper
				Teaspoon
			– fresh or canned	

Variation Slices of breadfruit and sweet-corn kernels also make good fritters.

Method
1. Drain fruit and sprinkle with flour.
2. Make a batter by beating the flour with egg and slowly add the milk until smooth.
3. Dip fruit, e.g. rings of pineapple or slices of banana, in this and fry on each side in hot oil.
4. Drain on kitchen paper and sprinkle with sugar, flavoured with cinnamon.

Serve warm with wedges of lemon or orange.

Citrus Sponge Pudding

'Magic' to make! A pudding which has a smooth citrus sauce running under a cake-like surface.

Quantity			Ingredients	Equipment
1	1	1	Lime or lemon	Baking dish – 375 ml/$\frac{3}{4}$ pt size
2	2	2	Eggs	Grater
3 oz	75 g	$\frac{1}{3}$ C	Sugar	Fruit squeezer
				Mixing bowls – 2
				Wooden spoon

Quantity			Ingredients	Equipment
2 oz	50 g	$\frac{1}{4}$ C	Margarine or butter	Whisk
2 oz	50 g	$\frac{1}{2}$ C	Flour	Teaspoon
$\frac{1}{4}$ tsp	1 ml	$\frac{1}{4}$ tsp	Baking powder	Measuring jug
$\frac{1}{2}$ pt	250 ml	1 C	Water	Tablespoon

Variation Use a tablespoon of cocoa instead of fruit flavouring to make a hot fudge pudding.

Method
1. Set the oven to 350°F and grease the baking dish.
2. Finely grate the rind and squeeze the juice from the lime or lemon.
3. Separate the eggs and beat together yolks with the sugar, margarine, flour, baking powder and water until smooth.
4. Whisk egg whites until stiff and fold carefully into the other ingredients using a tablespoon.
5. Pour into the dish and bake for 25–30 minutes.

Serve warm, sprinkled with confectioner's sugar and twists of sliced lemon or lime.

Crunchy Mango Crumble

A high-fibre dessert using an abundance of mangoes or any other seasonal fruits.

Quantity			Ingredients	Equipment
4 oz	100 g	1 C	Flour – wholewheat or all-purpose	Baking dish – 2 pt/ 1 litre size
2 oz	50 g	$\frac{2}{3}$ C	Porridge oats	Mixing bowl
3 oz	75 g	$\frac{1}{3}$ C	Margarine or butter	Tablespoon Teaspoon
3 oz	75 g	$\frac{1}{3}$ C	Brown sugar	Sharp knife
$\frac{1}{2}$ tsp	2 ml	$\frac{1}{2}$ tsp	Cinnamon	
1 oz	25 g	$\frac{1}{4}$ C	Chopped nuts	Cutting board
1 lb	450 g	2 C	Fruit, e.g. Mangoes, apples, plums	

Variation Use a can of fruit pie filling instead of fresh fruit.

Method
1. Set the oven to 350°F and grease the baking dish.
2. Place the flour into a mixing bowl and rub in the margarine using fingertips.
3. Stir the oats, half of the brown sugar, cinnamon and nuts into the flour mixture.
4. Wash and peel the fruit and cut into slices.
5. Place half of the fruit into the dish and sprinkle with the sugar. Cover with the remaining fruit.
6. Spread the crumble mixture on top and bake for 30 minutes, until golden brown.

Serve hot with custard sauce or cold with cream or evaporated milk.

Pineapple Upside-Down Cake

Using the creaming method of cake-making this popular dessert can be made using alternative fruits, such as sliced bananas, red-skinned apples, plums or guavas for the topping when turned over.

Quantity			Ingredients	Equipment
6 oz	150 g	$\frac{2}{3}$ C	Sugar	Baking pan – 9 ins/ 23 cm deep size
6 oz	150 g	$\frac{2}{3}$ C	Margarine or butter	Mixing bowl
3	3	3	Eggs	Wooden spoon
6 oz	150 g	$1\frac{1}{2}$ C	Flour	Sieve
2 tsp	10 ml	2 tsp	Baking powder	Tablespoon
$\frac{1}{2}$ tsp	2 ml	$\frac{1}{2}$ tsp	Vanilla flavouring	Teaspoon
1 oz	25 g	1 tblsp	Brown sugar or syrup	Can-opener Spatula
1	1	1	Pineapple slices – small can size	Serving plate
1 oz	25 g	1 tblsp	Pecan nuts	
Small amount			Cherries	

Variation Use sliced seasonal fresh fruit instead of canned.

Method
1. Set oven to 350°F and grease the baking pan.

2. Cream together the sugar and margarine in a bowl, using a wooden spoon.
3. Beat in the eggs with a little of the sifted flour.
4. Add the remaining flour and baking powder. If the consistency is stiff, add a small amount of the fruit juice from the can.
5. Sprinkle the brown sugar in the bottom of the baking pan and place the slices of fruit, cherries and nuts on top with some of the juice.
6. Pour cake mixture on top and bake for 40 minutes, until golden brown and firm to touch. Turn out on to a plate.

Serve warm or cold with pouring or whipped cream.

Ginger Lime Mousse

Tangy limes make enjoyable desserts in addition to refreshing drinks, marmalade, salad dressing and a seasoning for meat and fish.

Quantity			Ingredients	Equipment
1 oz	25 g	1 tblsp	Cornstarch	Serving dish with cover or individual size
$\frac{3}{4}$ pt	375 ml	$1\frac{1}{2}$ C	Milk	
4 oz	100 g	$\frac{1}{2}$ C	Sugar	
3	3	3	Eggs	Tablespoon
4	4	4	Limes	Measuring jug
1 tsp	5 ml	1 tsp	Ginger – fresh	Cooking pots – 2
$\frac{1}{2}$ oz	15 g	2 tblsp	Gelatine	Wooden spoon
Few drops			Green food colouring	Teaspoon
				Grater
				Sharp knife
				Cutting board
				Fruit Squeezer
				Mixing bowl
				Whisk

Variation Use chopped and seedless soursop, sapodillas or mangoes instead of limes.

Method

1. Mix cornstarch with a little milk. Beat in sugar and separated egg yolks.
2. Heat rest of the milk and stir into the mixture. Return to the cooking pot and stir until the mixture thickens. Take off the heat.

Food group 5: Fruits   75

3. Finely grate the 'zest' from the limes and add a small amount to the custard, with the finely chopped ginger and a few drops of green food colouring.
4. Squeeze the juice from the limes and dissolve the gelatine in this, over a gentle heat.
5. Whisk egg whites and fold into the mixture, with the gelatine. Spoon into the serving dish and chill.
   * Always dissolve gelatine in a small bowl or cup over a cooking pot of hot water.

Serve decorated with slices of lime and home-made gingerbread biscuits.

West Indian Trifle

The taste of the small sugar bananas is similar to that of strawberries, making wonderful banana custard or can be layered with cake, custard and cream to produce a special party dessert.

Quantity			Ingredients	Equipment
1	1	1	Sponge cake – 8 ins/20 cm round	Serving dish – large or individual size
2 tblsp	30 ml	2 tblsp	Jelly/jam, e.g. Raspberry, guava	Table knife
For the custard sauce:				
½ pt	250 ml	1 C	Milk	Cooking pot
2	2	2	Egg yolks	Measuring jug
1–2 oz	25–50 g	1–2 tblsp	Sugar	Tablespoon
1 tsp	5 ml	1 tsp	Cornstarch	Teaspoon
¼ tsp	1 ml	¼ tsp	Vanilla flavouring	Wooden spoon
4	4	4	Bananas – small size	Sharp knife
Small amount			Lime juice or liqueur	Cutting board
¼ pt	125 ml	½ C	Whipping cream – fresh, frozen or canned	Mixing bowl
To decorate			Almonds – flaked	Whisk
To decorate			Glacé cherries	Piping bag and nozzle

Variation Use canned raspberries or mixed fresh fruit instead of bananas and the egg whites to make a hot meringue topping instead of piped cream.

Method
1. Cut the sponge cake into pieces and spread with fruit jelly. Place in the serving dish.
2. To make the custard sauce: Heat the milk, leaving a small amount to blend the egg yolks, sugar and cornstarch together.
3. When the milk almost boils pour over the egg mixture and return to the cooking pot. Keep stirring on a low heat until the sauce thickens. Add the vanilla flavouring and allow to cool.
4. Peel and slice the bananas, sprinkle with lime juice and place on to the sponge. Cover with the custard.
5. Decorate with piped whipped cream, flaked almonds and chopped glacé cherries.

Serve chilled and decorated with chocolate flakes instead of glacé fruit.

Orange Banana Cake

Over-ripe bananas and all the other ingredients are easily combined in one bowl to make a heavy textured cake or bread, which often goes by the name of 'beat-up'!

Quantity			Ingredients	Equipment
8 oz	225 g	2 C	Flour	Loaf pan – 1 lb/ 450 g size
2 oz	50 g	$\frac{1}{4}$ C	Margarine or butter	Mixing bowl Sieve
2 oz	50 g	$\frac{1}{4}$ C	Sugar	Wooden spoon
1	1	1	Orange	Grater
2	2	2	Bananas	Fruit squeezer
1	1	1	Egg	Potato masher or fork
$\frac{1}{4}$ pt	125 ml	$\frac{1}{2}$ C	Milk	Measuring jug Tablespoon Spatula Cake skewer Cooling rack

Variation Use vanilla flavouring, spice, e.g. nutmeg or lemon instead of orange.

Method
1. Set the oven to 350°F and grease the loaf pan.
2. In a mixing bowl, rub the margarine into the flour using fingertips. Stir in the sugar.
3. Finely grate the rind and squeeze the juice from the orange.
4. Mash the bananas and add to the mixture with the orange.
5. Beat to a smooth consistency with the egg and milk. Place the mixture into the loaf pan.
6. Bake for 45 minutes – 1 hour until cooked. Test with a cake skewer, which should come out clean when inserted in the centre.

Serve sliced with butter or pieces of cheese. This also makes good banana sandwiches.

Mixed Fruit Marmalade

Mixed citrus fruit and sugar are boiled rapidly until thick to produce a jar of home-made preserve, tempting the production of a wide range of fruit jams, jellies and chutneys.

Quantity			Ingredients	Equipment
1	1	1	Grapefruit	Jars and covers
1	1	1	Orange	
2	2	2	Limes or lemons	Sharp knife
1	1	1	Pineapple – small size – fresh or canned	Cutting board
$2\frac{1}{2}$ pts	$1\frac{1}{4}$ l	5 C	Water	Muslin or fine cloth Preserving pan – large size
5 lbs	$4\frac{1}{2}$ kg	10 C	Sugar – granulated	Wooden spoon Thermometer Small plate

Variation Use small kumquat (or calamondin) for a delicate flavoured marmalade, omitting the pineapple and using 1 lb/450 g less sugar.

Method
1. Thinly peel the citrus rind and shred. Discard the pith, chop the fruit and put pips into a small piece of muslin cloth.
2. Peel and chop the pineapple.
3. Put the fruit, citrus peel and pips to soak in the water and set aside for at least 12 hours.
4. When soaked, place in a preserving pan, bring to the boil and simmer for about $1\frac{1}{2}$ hours (or pressure cook) until the peel is tender.
5. Remove the pips, stir in the sugar and boil until the mixture sets.
   * Test for setting: thick consistency on a cold plate or when the temperature is 221°F/105°C with a thermometer.
6. Pour into hot sterile jars, seal and label.

Serve with toasted home-made wholewheat bread or warmed pancakes.

FOOD GROUP 6:
Fats and oils

The fats we eat are usually added during cooking, e.g. margarine, butter, shortening, cooking oil and salt pork. 'Hidden' fats are found in many foods, e.g. meat (especially sausage), oily fish, nuts (coconut), avocado pear and salad dressings.

Fat is a more concentrated source of energy than carbohydrate. It is essential to have an energy supply for all the processes of living: breathing, keeping a constant body temperature and for blood circulation, as well as physical exercise. When the food we eat provides us with more energy than we need, the excess is converted to a store of fat in the body. Therefore a limited amount of fat should be eaten, as too much will become associated with weight problems and disease.

Use low fat alternative foods in recipes wherever possible, e.g. skimmed milk, low fat yogurt and cottage cheese. Polyunsaturated fats in the form of vegetable oil and margarine are preferable to using animal fats. You can also fry foods in only a small amount of cooking oil in a non-stick pan.

Although the recipes in this section should be eaten in moderation, remember that fats used in cookery do have some advantages. Nutritionally they do provide the body with energy, which is measured in units called kilocalories (or kilojoules), which are usually written as calories (joules). Include foods from this group when planning meals for certain groups, such as manual workers and growing children. Fats and oils also supply the body with the important vitamins A and D.

The use of fats often makes food very palatable and enables a person to stay satisfied for a longer period of time. They enable us to make dishes such as cakes and pastries which do require skills, but which should be eaten at the end of a meal and not instead of a regular meal.

Avocado Dip

A fat-enriched fruit which mixes well with other raw fruits, vegetables and seasonings in salads, spreads and appetisers.

Quantity			Ingredients	Equipment
1	1	1	Avocado – ripe	Serving bowl –
8 oz	250 ml	1 C	Sour cream – carton	small size with cover
				Mixing bowl
2	2	2	Garlic cloves or garlic salt	Sharp knife
Small amount				Cutting board
$\frac{1}{2}$	$\frac{1}{2}$	$\frac{1}{2}$	Lime	Tablespoon
Small amount			Black pepper	Potato masher
To taste			Hot sauce or chilli powder	Fruit squeezer
				Teaspoon
3	3	3	Bacon rashers	Frying pan

Variation Mix slices of peeled grapefruit and avocado for a starter.

Method
1. Cut the avocado in half and scoop out the flesh into a bowl.
2. Mash with the sour cream and finely chopped garlic cloves.
3. Add the juice of the lime and additional seasoning.
4. Broil/grill or lightly fry the bacon until crisp.
5. Place the dip into a serving bowl and garnish with crumbled bacon on top.

Serve with vegetable sticks, e.g. celery and carrot and some packet corn chips.

Salad Dressings

Fats and oils give an appetising and edible appeal to salads and fish dishes, resulting also in an increase in the calorific value. Use natural yogurt for a 'lite' dressing.

Quantity			Ingredients	Equipment
Mayonnaise:				
1	1	1	Egg	Container – small size with cover
1 tblsp	15 ml	1 tblsp	Lime juice	
$\frac{1}{2}$ tsp	2 ml	$\frac{1}{2}$ tsp	Mustard	Mixing bowl
$\frac{1}{2}$ tsp	2 ml	$\frac{1}{2}$ tsp	Sugar	Tablespoon
Small amount			Salt and pepper	Teaspoon
$\frac{1}{4}$ pt	125 ml	$\frac{1}{2}$ C	Salad oil	Measuring jug
				Electric blender or hand whisk

Food group 6: Fats and oils **81**

Quantity			Ingredients	Equipment

Mayonnaise variations:
1) Island dressing:

Quantity			Ingredients	Equipment
½ pt	250 ml	1 C	Mayonnaise	Sharp knife
1	1	1	Egg – hard boiled	Cutting board
¼	¼	¼	Sweet pepper	
1 tsp	5 ml	1 tsp	Chives or mixed herbs	
Small amount			Salt and pepper	
2 tblsp	30 ml	2 tblsp	Hot sauce	

2) Seafood sauce:

2 tblsp	30 ml	2 tblsp	Mayonnaise	
4 tblsp	60 ml	4 tblsp	Natural yogurt	
1 tblsp	15 ml	1 tblsp	Tomato paste	
1 tsp	5 ml	1 tsp	Soya sauce	
1 tsp	5 ml	1 tsp	Horseradish sauce	
Small amount			Garlic salt	
Small amount			Cayenne pepper	
2 tblsp	30 ml	2 tblsp	Lime juice	

Method
1. To make the mayonnaise use an electric blender or a hand whisk to combine the ingredients together. It is important to whisk in each drop of oil before adding the next.
   * To make the mayonnaise thicker slowly add more oil. To make the mayonnaise thinner add more lime juice or vinegar.
2. To make the island dressing and the seafood sauce mix together the mayonnaise with the other ingredients (chopped if necessary) and seasoning in a bowl.

Serve chilled with various vegetable salads or with seafood: crawfish, tuna or crab.

Pork Wheels

Sausage rolls or 'pigs in a blanket' made with flaky pastry make a very acceptable buffet party plate.

Quantity			Ingredients	Equipment
5 oz	125 g	⅔ C	Margarine or butter	Baking pan – flat size

Quantity			Ingredients	Equipment
8 oz	225 g	2 C	Flour	Mixing bowl
Small amount			Salt	Sieve
To mix			Water	Grater
8 oz	225 g	1 C	Sausage meat – minced or use hamburger	Measuring jug Table knife Rolling pin
2 tsp	10 ml	2 tsp	Tomato paste	Flour dredger
2 oz	50 g	¼ C	Cheese – grated	Teaspoon
Small amount			Pepper and mixed herbs	Tablespoon

Variation Use mashed canned sardines instead of pork meat for a filling.

Method
1. Set the oven to 400°F and grease the baking pan.
   * Place the fat into the freezer for 10 minutes to harden.
2. Sieve the flour and salt into a mixing bowl. Grate the cold fat into the flour. Stir in then slowly add the water to mix. Chill dough in the refrigerator.
3. Place the meat into the mixing bowl and mix in the other ingredients.
4. Roll out the pastry into an oblong and spread with the filling, almost to the edges.
5. Roll up the pastry away from the body, as for a 'jelly roll'. Cut into even slices.
6. Place on to the baking pan and bake for 20 minutes.

Serve warm with salad. This is also good packed up for lunch.

Tropical Slice

A rich French pastry recipe topped with a jelly and fresh coconut mixture and cut into 'fingers' makes a good fund-raiser!

Quantity			Ingredients	Equipment
6 oz	150 g	1½ C	Flour	Baking pan – flat size
3 oz	75 g	⅓ C	Butter or margarine	Mixing bowl
2 oz	50 g	¼ C	Sugar	Sieve
1	1	1	Egg – small size	Table knife

Quantity			Ingredients	Equipment
4 tblsp	60 ml	4 tblsp	Guava jelly	Rolling pin

For the topping:

3 oz	75 g	1 C	Coconut – grated	Flour dredger
4 oz	100 g	½ C	Sugar	Tablespoon
1	1	1	Egg	Bowl – small size
				Cooling rack

Variation Use an alternative flavoured fruit jam or jelly.

Method
1. Set the oven to 350°F and grease the baking pan.
2. Make the pastry by rubbing the butter into the flour, using the fingertips. Stir in the sugar and bind together with the egg.
3. Roll out the pastry into two 3 ins/7½ cm wide strips. Place on to the baking pan and pinch the outer edges.
4. Spread with the jelly. Mix together the ingredients for the topping in the mixing bowl and place over the surface.
5. Bake for 20–25 minutes. Cut into slices before cooling.

Serve with melted chocolate drizzled on the top.

Coconut Pie

The coconut produces oil which is used in the manufacture of margarine, and for a refreshing 'milk' drink, and the flesh which adds a distinctive flavour and texture in many dishes.

Quantity			Ingredients	Equipment
8 oz	225 g	2 C	Flour	Pie dish – 9 ins/ 23 cm size
4 oz	100 g	½ C	Margarine or butter	Mixing bowl
To mix			Water	Sieve

Filling:

5 oz	125 g	1½ C	Coconut – grated (approximately 1 fresh)	Measuring jug
1	1	1	Sweetened condensed milk – 14 oz/400 g can size	Table knife

Quantity			Ingredients	Equipment
2	2	2	Egg yolks	Rolling pin
1 tsp	5 ml	1 tsp	Vanilla flavouring	Flour dredger
$\frac{1}{2}$ tsp	2 ml	$\frac{1}{2}$ tsp	Nutmeg – grated	Cooking pot
$\frac{1}{2}$ tsp	2 ml	$\frac{1}{2}$ tsp	Cinnamon	Can-opener
Small amount			Lemon or orange – grated rind	Teaspoon
$\frac{1}{4}$ pt	125 ml	$\frac{1}{2}$ C	Water	Tablespoon

Variation For a soft sweet pastry add a little sugar and bind together with evaporated milk and water.

Method
1. Set the oven to 350°F and grease the pie dish.
2. Prepare the pastry by the rubbing-in method. Mix to a dough with cold water and roll out two-thirds of this to line the pie dish.
3. Combine all the ingredients together in a cooking pot for the pie filling and cook gently for about 10 minutes. Remove from the heat and cool.
4. Place the filling into the unbaked pastry case and dampen the edges.
5. Roll out the remaining pastry and cut into thin strips. Use to make a lattice cover for the pie. Bake for 35–40 minutes.

Serve cut into slices with pouring cream.

* A potato peeler can be used to remove the brown skin from the fresh coconut. The grated coconut freezes well in packages until required in a recipe.

Fudge Brownies

Individual dense rich chocolate cakes, especially good for kitchen helpers!

Quantity			Ingredients	Equipment
4 oz	100 g	$\frac{1}{2}$ C	Margarine or butter	Baking pan – 7 × 12 ins/18 × 30 cm size
8 oz	225 g	1 C	Brown sugar	Mixing bowl
2	2	2	Eggs	Wooden spoon
4 oz	100 g	1 C	Flour	Sieve
2 oz	50 g	$\frac{1}{4}$ C	Cocoa	Tablespoon
1 tsp	5 ml	1 tsp	Baking powder	Teaspoon
1 tblsp	15 ml	1 tblsp	Milk	Spatula

Food group 6: Fats and oils 85

Quantity	Ingredients	Equipment
To decorate	Frosting and chopped nuts	

Variation Add chopped pecan nuts to the mixture before baking.

Method
1. Set the oven to 350°F and grease the baking pan.
2. Cream the margarine and sugar together until pale and fluffy.
3. Slowly beat in the eggs with a little of the sieved flour.
4. Carefully fold in the remaining flour and sieved cocoa and baking powder.
5. Stir in the milk. Place in the baking pan and smooth the surface.
6. Bake for 30-35 minutes until firm to the touch.

Serve spread with chocolate or peppermint frosting and cut into squares.

Bounty Bars

Unfortunately high in calories, these cookie bars make a scrumptious sweet treat.

Quantity			Ingredients	Equipment
4 oz	100 g	½ C	Margarine or butter	Baking pan - 7 × 12 ins/18 × 30 cm
8 oz	150 g	2 C	Biscuits - graham cracker or digestive	Cooking pot
1	1	1	Sweetened condensed milk - 14 oz/400 g can size	Plastic bag Rolling pin
6 oz	150 g	1 C	Chocolate chips - semi-sweet	Wooden spoon
3 oz	75 g	1 C	Coconut - grated	Can-opener
4 oz	100 g	1 C	Chopped nuts	Tablespoon

Variation Add a few chopped glacé cherries to the mixture.

Method
1. Set the oven to 350°F and grease the baking pan.

2. Melt the margarine in a cooking pot and add the biscuit crumbs.
   * Make the crumbs by placing the biscuits in a plastic bag and using a rolling pin.
3. Press the crumbs into the baking pan and pour the condensed milk evenly over the top.
4. Top with the remaining ingredients, finishing with the nuts.
5. Bake for 25-30 minutes until beginning to brown. Cut into bars while still warm. Store in a covered container in the refrigerator.

Serve bars as a dessert with coffee.

Sweet Mince Pie

A Christmas sweetmeat, which can have suet (pork fat) or vegetable fat added and made into individual rich pastry pies. A good idea is to make the mincemeat in a preliminary lesson and bottle or store in the refrigerator, ready to use.

Quantity			Ingredients	Equipment
Mincemeat:				
4 oz	100 g	1 C	Raisins – dark	Mixing bowls – 2
4 oz	100 g	1 C	Raisins – golden	
4 oz	100 g	1 C	Currants	Tablespoon
4 oz	100 g	1 C	Dates – dried	
4 oz total amount	100 g	$\frac{1}{2}$ C	Cherries – glacé	Sharp knife
			Pineapple – crystallised	
			Citrus fruit peel – candied	Cutting board
2 oz	50 g	$\frac{1}{2}$ C	Almonds	
2 oz	50 g	$\frac{1}{4}$ C	Suet or vegetable margarine	Grater
2 oz	50 g	$\frac{1}{4}$ C	Brown sugar	Fruit squeezer
8 oz	225 g	1 C	Apples	
$\frac{1}{2}$ tsp	2 ml	$\frac{1}{2}$ tsp	Nutmeg – fresh – grated	Teaspoon
$\frac{1}{2}$	$\frac{1}{2}$	$\frac{1}{2}$	Lime	
$\frac{1}{2}$	$\frac{1}{2}$	$\frac{1}{2}$	Orange	
2 tblsp	30 ml	2 tblsp	Rum or brandy	

Quantity			Ingredients	Equipment
Rich pastry:				
8 oz	225 g	2 C	Flour	Cup cake baking pan
5 oz	125 g	$\frac{2}{3}$ C	Butter or margarine	Sieve Table knife
1 oz	25 g	1 tblsp	Sugar	Rolling pin
1	1	1	Egg yolk	Flour dredger
$\frac{1}{2}$ tsp	$\frac{1}{2}$ tsp	$\frac{1}{2}$ tsp	Lime juice	Pastry brush Biscuit cutters – 2
Small amount			Water	

Variation For speed use bought mincemeat. The rich pastry can also be used to make individual apple pies.

Method
1. Prepare the mincemeat by mixing all the ingredients together in a large bowl. Chop the dates, almonds and glacé fruit. Grate the apples, nutmeg and 'zest' from the lime and orange. Squeeze the juice also from the citrus fruit.
2. Set the oven to 425°F and grease the baking pans.
3. Prepare the pastry by rubbing the butter into the flour, using the fingertips. Add the sugar, egg yolk and lime juice. Bind together with a little water if necessary.
4. Roll out the pastry and cut into small rounds for the bases of the pies. Re-roll the pastry and cut out slightly smaller rounds for the lids. This quantity of pastry should make 12 pies.
5. Place the base rounds into the cup cake pan and spoon in the mincemeat filling. Brush the edges of the pastry with water. Place the pastry lids on top and seal the edges.
6. Make a small slit on the surface of each pie and brush with milk. Bake for 20–25 minutes until the pastry is firm and golden brown.

Serve sprinkled with sifted confectioner's sugar and with rum-flavoured butter cream.

Spicy Doughnuts

The thickened batter mix is shaped into hoops or twisted plaits and deep-fat fried to produce delicious 'donuts'.

Quantity			Ingredients	Equipment
8 oz	225 g	2 C	Flour	Serving plate
2 tsp	10 ml	2 tsp	Baking powder	Mixing bowl
Small amount			Salt	
1–2 oz	25–50 g	1–2 tbsp	Sugar	Teaspoon
1 oz	25 g	1 tblsp	Margarine or cooking oil	Sieve
1	1	1	Egg	Tablespoon
6–8 tblsp	90–120 ml	6–8 tblsp	Milk and water – mixed	Wooden spoon
For frying			Cooking oil	Rolling pin
				Flour dredger
$\frac{1}{4}$ tsp	1 ml	$\frac{1}{4}$ tsp	Cinnamon	Frying pan
$\frac{1}{4}$ tsp	1 ml	$\frac{1}{4}$ tsp	Nutmeg – grated	Draining spoon
Extra to coat			Sugar	Kitchen paper

Variation Make jelly/jam doughnuts by making a hollow in the centre of round doughnuts using the handle of a wooden spoon, fill and re-roll to cover.

Method
1. Sieve the dry ingredients into a mixing bowl. Mix in the sugar, melted margarine or cooking oil and egg.
2. Slowly add the milk and water and mix to a soft dough consistency. If the dough becomes too sticky cover and leave in the refrigerator for 10 minutes.
3. Roll out the dough on to a floured surface and cut into rounds. Alternatively use the hands to twist strands of the dough or make into small ball shapes.
4. Fry in hot cooking oil 3–4 ins/9 cm deep for 3–5 minutes.
   * The temperature should be 370–380°F/190°C with a thermometer.
5. Remove with a draining spoon and coat with the spices mixed with sugar on kitchen paper.

Serve some doughnuts plain, rolled in toasted coconut or dipped in melted chocolate.

Oat Nut Biscuits

Store oatmeal cereal in a sealed container and use it to make an inexpensive breakfast porridge, puddings, bread and other bakes.

Food group 6: Fats and oils

Quantity			Ingredients	Equipment
4 oz	100 g	½ C	Margarine or butter	Baking pans – flat size
4 oz	100 g	½ C	Brown sugar	Mixing bowl
4 oz	100 g	½ C	Granulated sugar	Wooden spoon
1	1	1	Egg	Teaspoon
4 oz	100 g	1 C	Flour	Grater
1 tsp	5 ml	1 tsp	Baking powder	Fork
1 tsp	5 ml	1 tsp	Baking soda	Cooling rack
2 oz	100 g	1 C	Rolled oats	
3 oz	75 g	1 C	Coconut – grated	
1 tsp	5 ml	1 tsp	Vanilla flavouring	

Variation Add a tablespoon of chopped nuts, raisins, chocolate chips or grated apple to the mixture.

Method
1. Set the oven to 350°F and grease the baking pans.
2. Cream together the margarine and sugars in a mixing bowl with a wooden spoon.
3. Mix in the egg and the dry ingredients.
4. Add the vanilla flavouring and use the hands to make into a dough.
5. Place small ball shapes on to the baking pan, spacing well apart, and press down with a fork.
6. Bake for 10–15 minutes until golden brown. Allow to harden on the baking pan.

Serve also made into jumbo cookies or bars, by spreading the mixture into a baking pan.

COOKERY TERMS

To assist you with understanding the recipes, here is an explanation of some of the cookery terms used.

APPETISER – a food served to stimulate the appetite, as the first course of a meal.
To *BASTE* – to spoon the juices over meat when cooking to keep it moist.
BATTER – a mixture of flour and liquid, usually egg, beaten until smooth.
To *BEAT* – to mix ingredients well together incorporating air.
To *BIND* – to add liquid, usually milk or egg, to hold ingredients together.
To *BLEND* – to combine a liquid and powder, for a smooth paste.
To *CHILL* – to place in the refrigerator until cold.
To *CHOP* – to cut food into small pieces.
To *COAT* – to cover food with a protective layer before cooking.
CONSISTENCY – the firmness of a mixture, which can be dropping, soft or a stiff dough.
To *CREAM* – to beat together fat and sugar until light and fluffy, using a wooden spoon or mixer.
To *DECORATE* – to make sweet foods look attractive.
To *DICE* – to cut food into small cubes.
To *DOT* – to scatter small pieces of fat over the surface of food.
DOUGH – a mixture of flour and liquid stiff enough to knead.
To *DRAIN* – to remove liquid from canned or other food.
To *FLAKE* – to break food into pieces using a fork.
To *FILLET* – to remove the bone from a cut of meat or fish.
To *FOLD IN* – to add a dry ingredient carefully to a mixture with a 'figure of eight' movement, using a tablespoon.
To *FROST* – to decorate with icing.
To *GARNISH* – to decorate savoury foods.
To *GLAZE* – to brush the surface of baked products with beaten egg or milk to give a shiny appearance.
To *GRATE* – to cut food into small pieces by rubbing against a sharp grater.

To *GREASE* – to spread the inside of a baking pan lightly with fat to prevent food from sticking.

To *KNEAD* – to make a stretchy dough by a folding and pressing action of the hands. 'Knocked-back' refers to re-kneading.

To *LINE* – to cut foil or wax paper to fit the baking pan and grease before using.

To *MARINATE* – to add seasoning and lime juice to meat or fish before cooking, to give flavour and tenderise.

To *MASH* – to remove lumps from food by using a potato masher or fork.

To *MIX* – to combine two or more ingredients together.

To *PEEL* – to remove outer skin from fruit or vegetables.

To *PIPE* – to decorate with cream or frosting using a bag and nozzle.

To *PROVE* – to rise dough before baking as a result of the action of yeast.

To *PURÉE* – to make a smooth pulp by using a sieve or blender.

To *ROLL OUT* – to use a rolling pin on a floured surface, with light strokes in one direction only.

ROUX – an equal quantity of fat and flour used to thicken a sauce.

To *RUB IN* – to incorporate fat into flour using fingertips until the mixture resembles breadcrumbs.

To *SEAL* – to join two pastry edges together firmly by moistening them with water or egg and fluting them if desired.

To *SEASON* – to add flavouring to food using herbs and spices.

To *SHAPE* – to make cut dough into a desired pattern and size.

To *SHRED* – to cut or slice food very finely.

To *SIEVE/SIFT* – to pass food through a sieve to make finer.

To *SOAK* – to cover food with liquid before using.

To *SPRINKLE* – dredge or coat food lightly with flour or sugar.

To *SQUEEZE* – to remove juice from citrus fruit using a lemon squeezer.

To *STERILISE* – to destroy micro-organisms by using heat.

STOCK – the liquid from cooking vegetables, meat (bones) or fish, used to make soups and sauces.

To *STRAIN* – to remove liquid from food using a sieve.

To *TOSS* – to mix ingredients lightly together without crushing.

To *WHISK* – to whip the ingredients rapidly, so that air is enclosed and the mixture will thicken and be light when cooked.

ZEST – the thinly peeled rind of citrus fruit.

METHODS OF COOKING

Food is cooked to make it safer to eat and improve the keeping quality. Cooking also makes food tender, easily digestible and improves the flavour, colour and texture, making it more appetising.
Heat (energy) is used to cook food by:
1. Conduction – passage of heat through a solid object, e.g. boiling.
2. Radiation – heat directed straight on to food, e.g. broiling or grilling.
3. Convection – circular movement of hot air, e.g. baking.

The methods of cooking are usually divided into:
A. Moist methods – boiling, steaming, stewing.
B. Dry Methods – baking, broiling or grilling.
C. Hot fat methods – frying, roasting.

Other methods of classification include divisions according to their advantages and disadvantages: slow and quick methods of cooking and types of food cooked by each method, such as lean and tough cuts of meat.

The following methods of cooking are used in the recipes:
BAKING – cooking food in dry heat in an oven, in a greased baking pan until brown.
BARBECUING – cooking food outside on a grate over a charcoal fire.
BOILING – cooking food in rapidly boiling liquid at 212°F/100°C.
BRAISING – frying the food, then adding a small amount of water and cooking in a covered dish inside the oven.
BROILING/GRILLING – cooking food directly below the source of heat.
BROWNING – placing the food under direct heat to toast the surface.
  * If the food is browning too quickly cover with aluminium foil wrap. Foil is also useful to cover food for carrying home.
CASSEROLING – cooking a stew-type meal in a covered dish inside the oven.

Methods of cooking

DOUBLE-BOILING – cooking food in a protective bowl surrounded by hot water.

FRYING – cooking food in hot fat or oil at 370–380°F/180°C.
  i) Deep frying – cooking food in plenty of fat in a deep pot.
  ii) Dry frying – cooking fatty foods with no extra fat added.
  iii) Shallow frying – cooking food in a small amount of fat, just covering the bottom of the pan.

MELTING – heating a food, so that it changes from a solid to a liquid.

MICROWAVE COOKING – cooking by electromagnetic radiation. Some of the recipes in the book can be adapted for use with a microwave oven.

PAR-BOILING – cooking food in boiling water and removing before cooking is complete, to continue cooking by another method.

POACHING – cooking food covered in hot liquid.

POT-ROASTING – cooking meat in a small amount of fat in a covered pot.

PRE-HEATING – setting the oven at the required temperature before the dish is fully prepared.

PRESSURE-COOKING – cooking food in a special covered pot under increased pressure, which increases the temperature of the liquid above boiling point.

ROASTING – cooking food in shallow hot fat in the oven, or on a spit over an open fire.

SAUTÉING – frying food gently in a small amount of melted hot fat.

SIMMERING – cooking food in a liquid that is just below boiling point, at 185°F/85°C.

STEAMING – cooking in steam rising from boiling water, where the food is placed above in a container with holes, in a covered bowl inside a cooking pot or in between two plates.

STEWING – cooking food in a small amount of simmering liquid in a pot on the stove or in the oven.

When baked items, e.g. bread, cakes and biscuits, have completed cooking, test by one of the following methods:
1. Cake skewer – which should come out clean when inserted in the centre.
2. Finger – pressed lightly – the cooked item should be firm to touch.
3. Colour – golden brown and good smell.
4. Shrinkage – away from the side of the baking pan.

MEAL PLANNING

There are some important rules to follow when planning meals.
1. Plan the meal according to the needs of the individual family so that it can be adapted as far as possible to suit everyone, for example a family perhaps consisting of an elderly person, small children, someone involved in hard manual work outside and an expectant mother. Each individual person will have different likes, dislikes and appetite according to his/her energy requirement
2. Ensure that the meal is well balanced, which can be achieved by combining foods from the six food groups to provide the essential nutrients. Make sure you include some fresh fruit and vegetables and protein (meat, fish, cheese, eggs, peas and beans) in the day's meals, and not too many sweet or fatty foods.
3. Relate what is chosen to the amount of money you have to spend. Make careful use of left-over food and economical cuts of meat. Preserve fresh fruit in season whenever possible, e.g. making jam and marmalade, bottling fruit and freezing for future use when prices are expensive. Become a good shopper by reading labels, checking dates and weight of food and comparing prices.
4. Work out how much time you have for preparing and cooking a meal, making use of labour saving equipment if possible, e.g. electric mixer/blender, pressure cooker and canned, packet and frozen 'convenience' foods, which can add variety to meals.
5. Make full use of the oven, being economical with the fuel supply. If you have the oven switched on, think of 'batch-baking' – baking double the amount or several dishes at once, which can be frozen and eaten another time. The hot weather may determine the type of food served, e.g. salad, cold dessert and beverage.
6. The meal may be for a special occasion, in which case three courses should be included: appetiser, main dish with salad or vegetables and dessert. If the main dish is very substantial

the other courses should be lighter. Choose a varying range of colours, textures and flavours for the meal. Avoid serving two pastry dishes or using the same ingredient many times, e.g. pineapple. You may also want to show your skill as a cook if you are entertaining guests and further decorate or garnish the dishes, making a good presentation of food.

* These rules will also help you when writing the *Reasons for Choice* section in your examination meal plan.

MENU IDEAS

A menu is a list of foods eaten at a particular meal. To assist you when planning meals, here are some ways of including foods from each of the six food groups in the day's meals. Remember that some dishes will be overlapping more than one food group.

Cream of Pumpkin Soup
Corn Bread Muffins

Fish Fingers
Pigeon Peas & Rice

Pineapple Upside-Down Cake

Lentil Curry Spread
Savoury Crackers

Jambalaya Rice
Sunshine Salad

Simple Cheesecake & Fruit

Carrot Pea Soup

Meat Loaf
Red Cabbage Slaw

Lattice Fruit Pie
Home-made Lemonade

Three Bean Salad

BBQ Ribs/Chicken
Curried Plantain

Frosted Pumpkin Cake
Chilled Orange Juice

Fish Chowder
Johnny Cake

Cabbage Casserole

Fresh Fruit Cocktail
Peanut Butter Cookies

Egg-Plant Bake
Cheese Shortbread

Pineapple Pork Chops

Fudge Brownies
Natural Yogurt

Pawpaw Nectar Pulse Burgers Broccoli & Cheese Sauce Raisin Bread Pudding *LV	Green Salad & Dressing Vegetable Hot-Pot Wholewheat Bread Crunchy Mango Crumble *SV

* LV - Menu suitable for a lacto-vegetarian.
* SV - Menu suitable for a strict vegetarian.

Menus suitable for 1 day

BREAKFAST Sugared Fruit Fritters or Boiled Egg with Wholewheat Bread & Mixed Fruit Marmalade Fresh Fruit Juice	**PACKED LUNCH** Hottie Patties or Cheese Pizza with Salad Orange Banana Cake Iced Tea

DINNER
Shell Fish Salad
Chilli Sauce with
Rice or Pasta
Chocolate Pudding Pie
Coffee

RECIPE INDEX

Animal Foods – Recipes 44
Avocado Dip 79

Baked Grouper Fish 49
Bakes –
   Basic Bean 28
   Egg-Plant 35
Batters –
   Spicy Doughnuts 87
   Sugared Fruit Fritters 70
   Tossed Pancakes 59
BBQ Ribs or Chicken 45
Beans –
   Basic Bean Bake 28
   Pulse Burgers 27
   Three Bean Salad 25
Bounty Bars 85
Biscuits – Oat Nut 88
Breads –
   Corn Bread Muffins 16
   Johnny Cake 15
   Orange Banana Cake 76
   Wholewheat Bread 13
Broccoli and Cheese Sauce 34

Cabbage Casserole 37
Cakes –
   Cup Cakes 53
   Frosted Pumpkin Cake 43
   Fudge Brownies 84
   Jelly Roll 55
   Johnny Cake 15
   Orange Banana Cake 76
   Pineapple Upside-
     Down Cake 73
   Simple Cheese-Cake 66

Callaloo 38
Caribbean Quiche 58
Carrot Pea Soup 30
Casseroles –
   Cabbage Casserole 37
   Vegetable Hot-Pot 23
   Liver Casserole 46
Cereals – Recipes 13
Cheese –
   Broccoli and Cheese
     Sauce 34
   Cheese and Ham Roll-Ups 63
   Cheese Pizza 65
   Cheese Shortbread 67
   Cheesy Staple Pie 18
   Lemon Cheese 56
   Macaroni Cheese 64
   Simple Cheese-Cake 66
Cheese and Ham Roll-Ups 63
Cheese-Cake – Simple 66
Cheese Pizza 65
Cheese Shortbread 67
Cheesy Staple Pie 18
Chilli Sauce 26
Chocolate Pudding Pie 60
Chowder – Fish 53
Citrus Sponge Pudding 71
Coconut Pie 83
Cooked Christophene 36
Cookies –
   Cheese Shortbread 67
   Peanut Butter 33
   Oat Nut Biscuits 88
Corn Bread Muffins 16
Cream of Pumpkin Soup 39
Crunchy Mango Crumble 72

Recipe index

Cup Cakes 53
Curries –
   Curried Plantain 20
   Lentil Curry Spread 32
   Custard – West Indian
          Trifle 75

Dark Green Leafy/Yellow
   Vegetables – Recipes 34
Dips –
   Avocado 79
   Lentil Curry Spread 32
Doughnuts –Spicy 87

Egg-Plant Bake 35
Eggs – Recipes 53

Fats and Oils – Recipes 79
Fish –
   Baked Grouper 49
   Callaloo 38
   Fish Chowder 53
   Fish Fingers 51
   Island Fish Pie 50
   Shell Fish Salad 52
Fresh Fruit Cocktail 68
Frosted Pumpkin Cake 43
Fruit –
   Fresh Fruit Cocktail 68
   Lattice Fruit Pie 69
   Mixed Fruit Marmalade 77
   Sugared Fruit Fritters 70
Fudge Brownies 84

Gelatine –
   Ginger Lime Mousse 74

Hottie Patties 47

Island Fish Pie 50

Jambalaya Rice 21
Jelly Roll 55
Johnny Cake 15

Lattice Fruit Pie 69
Legumes – Recipes 23
Lemon Cheese 56
Lentil Curry Spread 32
Liver Casserole 46

Macaroni –
   Cheese 64
   Savoury 19
Marmalade – Mixed Fruit 77
Meat –
   BBQ Ribs or Chicken 45
   Hottie Patties 47
   Liver Casserole 46
   Meat Loaf 44
   Pineapple Pork Chops 48
   Pork Wheels 81
Milk – Recipes 58
Mixed Fruit Marmalade 77
Muffins – Corn Bread 16

Natural Yogurt 62
Nuts –
   Nut Roast 31
   Oat Nut Biscuits 88
   Peanut Butter Cookies 33

Oat Nut Biscuits 88
Offal – Liver Casserole 46
Orange Banana Cake 76
Okra and Tomatoes 42

Pancakes – Tossed 59
Pasta –
   Savoury Macaroni 19
   Macaroni Cheese 64
Pastry –
   Caribbean Quiche 58
   Coconut Pie 83
   Lattice Fruit Pie 69
   Sweet Mince Pie 86
   Tropical Slice 82
Pawpaw Nectar 61

Peanut Butter Cookies 33
Peas –
  Carrot Pea Soup 30
  Pigeon Peas and Rice 29
Pies –
  Cheesy Staple Pie 18
  Chocolate Pudding Pie 66
  Coconut Pie 83
  Island Fish Pie 50
  Lattice Fruit Pie 69
  Sweet Mince Pie 86
Pineapple Pork Chops 48
Pineapple Upside-Down
  Cake 73
Pigeon Peas and Rice 29
Pizza – Cheese 65
Pork Wheels 81
Potato Salad 56
Poultry –
  BBQ Chicken 45
  Chicken Salad 52
Preserves –
  Lemon Cheese 56
  Mincemeat 86
  Mixed Fruit Marmalade 77
Puddings –
  Chocolate Pudding Pie 60
  Citrus Sponge Pudding 71
  Raisin Bread Pudding 14
  Sweet Potato Pone 17
Pulse Burgers 27
Pulses – Recipes 23

Raisin Bread Pudding 14
Red Cabbage Slaw 40
Rice –
  Jambalaya 21
  Pigeon Peas and Rice 29

Salads –
  Potato Salad 56
  Red Cabbage Slaw 40
  Sunshine Salad 41

Three Bean Salad 25
Salad Dressings – 80
  Coconut Cream 41
  Island Dressing 81
  Mayonnaise 80
  Seafood Sauce 81
Sauces –
  Broccoli and Cheese Sauce 34
  Cheese and Ham Roll-Ups 63
  Cheesy Staple Pie 18
  Chilli Sauce 26
  Custard 75
Savoury Macaroni 19
Shell Fish Salad 52
Simple Cheese-Cake 66
Soups –
  Callaloo 38
  Carrot Pea Soup 30
  Cream of Pumpkin Soup 39
  Fish Chowder 53
Spicy Doughnuts 87
Staples – Recipes 13
Stuffed Breadfruit 20
Sugared Fruit Fritters 70
Sunshine Salad 41
Supper Sandwiches 57
Sweet Mince Pie 86
Sweet Potato Pone 17

Three Bean Salad 25
Tossed Pancakes 59
Trifle – West Indian 75
Tropical Slice 82

Vegetables – Recipes 34
Vegetable Hot-Pot 23

West Indian Trifle 75
Wholewheat Bread 13

Yeast Mixtures –
  Wholewheat Bread 13
Yogurt – Natural 62